MATHEMATICS FOR INTRODUCTORY STATISTICS

MATHEMATICS FOR INTRODUCTORY STATISTICS

A Programmed Review

ANDREW R. BAGGALEY

Graduate School of Education
University of Pennsylvania
Philadelphia, Pennsylvania 19104

John Wiley & Sons, Inc.

New York • London • Sydney • Toronto

10 9 8 7 6 5 4 3 2 1

Library of Congress Catalog Card Number: 69-19103
Cloth: SBN 471 04007 X Paper: SBN 471 04008 8
Printed in the United States of America

TO MY FATHER,

WALTER BAGGALEY

Preface

Throughout my years of teaching statistics I have noticed that many of the difficulties encountered by students develop because their basic mathematical skills have become "rusty." Many students who undertake to learn statistics have not taken any formal courses in mathematics for five, ten, or even twenty-five years! This is particularly true of school teachers (other than mathematics or science teachers) who are returning to universities for graduate work. Actually most statistics courses require no more mathematics than is offered in the traditional two years of algebra given at either the high school or college freshman level. I have found that all but a few of my statistics students have been exposed to the requisite mathematical processes but that for many the memory of these processes has faded because of disuse during the intervening time.

The problem then became one of bringing these students back to the necessary level of mathematical sophistication in the period of a few weeks. Some of the available review books included topics (such as logarithms) which, although desirable, are not absolutely necessary for understanding introductory statistics. Therefore I undertook to construct a condensed review of the bare minimum of necessary processes. The techniques of programmed instruction afforded an opportunity for this goal to be attained with maximal efficiency. It should be emphasized, however, that this is a "review"; it is doubtful that a person with *no* previous exposure to these topics could learn them from "scratch."

The review is in four sections. Chapter I on algebra is basic in that some of the material is used in the development of the later sections. In Chapter II, after a review of the process of plotting points on graphs, the concepts of "intercept" and "slope" are introduced. The instructor may wish to develop them further. In Chapter III on extraction of the square root an algebraic rationale is offered. The more mathematically apt students will probably find that this rationale will aid their understanding of the steps involved. However, the process of extraction itself is self-contained in the sense that the other students should be able to obtain correct square roots even if they do not understand the "why" of the particular steps they are carrying out. Chapter IV, on use of the summation operator and summation laws does not constitute a "review" in the same sense as the earlier three parts. These topics at present are included in the regular mathematics courses of only a few high schools and colleges. However, the instructor of introductory statistics may wish to devote his class time to consideration of material more directly involved in statistics itself; the material in Chapter IV will help him to accomplish this aim.

All four parts of the review have been tested on university students. Their comments have been most helpful in revision of the parts. The following are estimates of the time the reader will need to work through each part:

75 minutes for algebra,
25 minutes for plotting points and linear equations on graphs,
45 minutes for extraction of the square root,
45 minutes for use of the summation operator and summation Laws.

Andrew R. Baggaley

Wyncote, Pennsylvania

Instructions to the Teacher

I have tried to write these materials so that they can be used for a variety of purposes. However, I shall indicate a way of using them that would seem to fit most introductory courses in applied statistics, that is, courses in which a knowledge of calculus is not assumed.

The first chapter, Algebra, should be read by the students as soon as possible. The manipulation of linear equations is a skill that many students have forgotten, and unless they recover it within a couple of weeks they are likely to find the material in statistics to be rough going.

The second chapter, Plotting Points and Linear Equations on Graphs, is probably best introduced just before consideration of the topic of correlation. The concepts of slope and intercept can then be applied directly to the constants in the regression equation.

Instructors of statistics differ on the question how to treat the problem of extracting the square root. If desk calculators are available, some teachers prefer to use them for this purpose. Some instructors teach their students to read a table of squares backward. The third chapter, Extraction of the Square Root, should be helpful to those teachers who wish the student to be able to obtain a square root with the use of only pencil and paper. The algebraic rationale is "icing on the cake" in the sense that the computational steps do not require it.

Statistics teachers also differ on how early they expect their students to be able to derive formulas. The three summation laws are particularly useful in the derivation of "raw-score" computational formulas. Thus they could be used as early as the concept of standard deviation. They would seem to be obligatory by the time that the topics of analysis of variance and multiple correlation have been reached.

I suggest that the teacher *not* ask the students to do all four chapters at the outset of the statistics course. Some of the weaker students have a short attention span for symbolic manipulation and tend to become discouraged unless they can see an immediate application for their labors.

In terms of objectives, it is expected that the student, upon completion of each chapter, will be able to:

Chapter 1

Express the products and quotients of cardinal numbers and unknowns in the standard algebraic form.

Recognize and generate equivalent fractions.

Add, subtract, multiply, and divide fractions.

Multiply and divide decimals (addition and subtraction of decimals are assumed to be already known.)

Add, subtract, multiply, and divide signed numbers.

Carry out in correct sequence the operations involved in expressions that contain parentheses and brackets.

Add, subtract, and multiply by zero, and divide into zero.

Multiply and divide square root radicals, and correctly introduce a factor under the square root radical.

Chapter 2

Use correctly the terminology referring to two-dimensional graphs.

Plot points on a two-dimensional graph.

Plot a linear equation on a two-dimensional graph.

Determine the y intercept and slope of a linear equation.

Chapter 3

Extract the square root of a number, accurate to a designated number of significant figures.

Chapter 4

Use the summation operator to express the sum of a series of scores.

Use the summation operator to express the sum of a series of squares of scores.

Use the three summation laws to change algebraic expressions into a form more convenient for statistical computations.

Instructions to the Reader

The material in this book consists of "frames" that present explanatory material followed by problems. The answer to each problem is also given. Notice the masking shield that is provided. To learn how to use this programmed book turn to the first page of the section on algebra. Notice that three sentences of explanatory material are given and then a sentence that presents a problem, namely, the product of 4 and 5. A blank is provided for your answer to this problem. Below the line is the correct answer, 20.

The purpose of the masking shield is to cover the answer to each problem until you have made your written response. After each response slide the shield down the page until it reaches the solid line that separates the frames. Then check your response against the answer that has just been uncovered. The program has been written in such small steps that about 80 to 90% of your responses should be correct. However, if your response to a problem is incorrect, you should try to determine how you went wrong *before* going on to the next frame. Check back to the earlier explanatory material if necessary. For purposes of later review it is helpful to indicate wrong responses by writing a large X in the margin rather than erasing the wrong response.

The blank for your response does not *always* appear at the end of the frame. Furthermore, some frames contain more than one blank to be filled in. You should *not* work on this material as though you were taking a speed test. One of the main advantages of programmed instructional material is that it allows each reader to work at his own pace. Thorough comprehension is the aim! At times you may feel tempted to "peek" at the answer before making your response to the problem. Try to resist this impulse because it would hinder your ultimate comprehension of the material.

At the end of each chapter is a review test in which you can evaluate your mastery of the preceding material.

Contents

MATHEMATICS FOR INTRODUCTORY STATISTICS

Chapter I. Algebra

OBJECTIVES

Upon completion of this chapter, the reader should be able to do the following:

1. Express the products and quotients of cardinal numbers and unknowns in the standard algebraic form.

2. Recognize and generate equivalent fractions.

3. Add, subtract, multiply, and divide fractions.

4. Multiply and divide decimals (addition and subtraction of decimals are assumed to be already known).

5. Add, subtract, multiply, and divide signed numbers.

6. Carry out in correct sequence the operations involved in expressions that contain parentheses and brackets.

7. Add, subtract, and multiply by zero, and divide into zero.

8. Multiply and divide square root radicals, and correctly introduce a factor under the square root radical.

9. Solve linear equations in one unknown.

BASIC TERMS AND NOTATION

Because the symbol x is used in algebra to indicate an unknown quantity, we shall *not* use it as a symbol for the operation of multiplication. Instead, we can use parentheses; for example 3(2) = 6. Thus 4(5) = __________.

- - - - - - - - - - - - - - -

20

If we want to express the product of 7 times 3 times 8 in this way, we can write ____________ or ____________ .

- - - - - - - - - - - - - - -

7(3)(8) [or] **(7)(3)(8)**

The quantities that are multiplied by each other are sometimes called "factors." Thus in the product 5(23)(11) the factors are ____________________________ .

- - - - - - - - - - - - - - -

5, 23, and 11

If one or more of the factors is an unknown, a simpler way of expressing a product is to write the factors directly adjacent to each other; for example, 9 times y can be expressed as $9y$. 38 times z can be expressed as __________.

- - - - - - - - - - - - - - -

38z

This method of expression is particularly helpful when we wish to show the product of more than one unknown. For example, 5 times w times x can be written as $5wx$. In the same manner, 46 times b times c can be written __________.

- - - - - - - - - - - - - - -

46bc

In representing the product of more than one number with an unknown or unknowns, we usually multiply the numbers first; for example, 3 times 8 times y is written as $24y$. Thus 11 times 4 times v can be written __________.

- - - - - - - - - - - - - - -

44v

---

4 times 15 times 3 times y times z is usually expressed as ______________.

- - - - - - - - - - - - - - -

180*yz*

In ordinary algebra, the operation of multiplication is "commutative." This means that the value of a product of factors is the same, no matter in which order the factors are arranged; for example, 4 times 9 equals 36; likewise 9 times 4 equals 36. Also, cw is algebraically equivalent to wc. Thus zx is equivalent to ______________.

- - - - - - - - - - - - - - -

xz

The fact that $13Af$ is equivalent to $13fA$ illustrates the ______________ property of ordinary algebra.

- - - - - - - - - - - - - - -

commutative

It is conventional in a product of numbers and letters to write the number or product of numbers first; for example, 6 times d times 9 is expressed as $54d$. J times 3 times q times 2 is expressed as ______________.

- - - - - - - - - - - - - - -

$6Jq$ [or] **$6qJ$**

In algebra the operation of division is usually indicated by use of a bar; for example, 53 divided by 6 is expressed as $\frac{53}{6}$.

In a similar fashion 4 divided by 7 can be expressed as ______,

and $25H$ divided by $3n$ can be expressed as __________.

- - - - - - - - - - - - - - -

$\frac{4}{7}$; $\frac{25H}{3n}$

The quantity 2 times 9 times L times x, divided by the quantity 13 times a, can be expressed as $\frac{18Lx}{13a}$. The quantity 23 times 5 times g times M, divided by the quantity 4 times w, can be expressed ______________________.

- - - - - - - - - - - - - - -

$\frac{115gM}{4w}$

The quantity, e times 7, divided by the quantity, 6 times Q times 4, is expressed as $\frac{7e}{24Q}$. The quantity 25 times x times w times 3, divided by the quantity A times 11, is expressed as

______________________________.

- - - - - - - - - - - - - - -

$\frac{75xw}{11A}$ [or] $\frac{75wx}{11A}$

EQUIVALENT FRACTIONS

The numerator and denominator of a fraction may each be multiplied by the same number or letter without changing the value of the fraction; for example, if we take the fraction $\frac{3}{5}$ and multiply the numerator and denominator each by 2, we will obtain the fraction $\frac{6}{10}$, which is algebraically equivalent to $\frac{3}{5}$. If we take the fraction $\frac{14}{9}$ and multiply the numerator and denominator each by 3,

we will obtain the equivalent fraction ______________.

- - - - - - - - - - - - - - -

$\mathbf{\frac{42}{27}}$

If we take the fraction $\frac{28}{17}$ and multiply the numerator and denominator each by z, we will obtain the equivalent fraction

______________.

- - - - - - - - - - - - - - -

$\mathbf{\frac{28z}{17z}}$

If we take the fraction $\frac{3x}{8y}$ and multiply the numerator and denominator by x, we will obtain the equivalent fraction $\frac{3x^2}{8xy}$. If we take the fraction $\frac{7F}{5m}$ and multiply the numerator and denominator by m, we will obtain the equivalent fraction ____________.

- - - - - - - - - - - - - - -

$\frac{7Fm}{5m^2}$

If we take the fraction $\frac{11uz}{8x}$ and multiply the numerator and denominator by $4u$, we will obtain the equivalent fraction

____________.

- - - - - - - - - - - - - - -

$\frac{44u^2z}{32ux}$

Likewise, if we *divide* the numerator and denominator of a fraction by the same number or letter, the resulting fraction is equivalent to the original fraction. In fact, this is the method by which we "reduce" fractions to their simplest form; for example, if we take the fraction $\frac{14}{6}$ and divide the numerator and denominator each by 2, we will have the equivalent fraction $\frac{7}{3}$. If we take the fraction $\frac{75}{50}$ and divide the numerator and denominator by 25, we will have the equivalent fraction ______________.

\- - - - - - - - - - - - - - -

$\mathbf{\frac{3}{2}}$

If we take the fraction $\frac{12ax}{20ay}$ and divide the numerator and denominator by $4a$, we will have the equivalent fraction $\frac{3x}{5y}$. If we take the fraction $\frac{35Gmw^2z}{14Gpw}$ and divide the numerator and denominator by $7Gw$, we will have the equivalent fraction ______________.

\- - - - - - - - - - - - - - -

$\mathbf{\frac{5mwz}{2p}}$

On the other hand, if we *add* or *subtract* the same number or letter from the numerator and denominator of a fraction, in general the value of the resulting fraction will be different; for example, if we take the fraction $\frac{3}{4}$ and add 1 to the numerator and denominator, we will have the fraction $\frac{4}{5}$, which is greater than $\frac{3}{4}$. Likewise, if we take the fraction $\frac{a}{b}$ and subtract 7 from the numerator and denominator, we will have the fraction ____________, which is *not* in general equal to $\frac{a}{b}$.

- - - - - - - - - - - - - - -

$$\frac{a-7}{b-7}$$

ADDITION AND SUBTRACTION OF FRACTIONS

We cannot add or subtract fractions unless they have the same denominator. If they do not, we must apply multipliers to one or both of the fractions so that they will have a common denominator; for example, we cannot perform the operation $\frac{2}{7} + \frac{3}{5}$ directly. However, we can multiply *both* the numerator and denominator of $\frac{2}{7}$ by 5, thus obtaining the equivalent fraction

__________________.

\- - - - - - - - - - - - - - -

$\frac{10}{35}$

Likewise, we can multiply both the numerator and denominator of $\frac{3}{5}$ by 7, thus obtaining the equivalent fraction ______________.

\- - - - - - - - - - - - - - -

$\frac{21}{35}$

Now we have two fractions, $\frac{10}{35}$ and $\frac{21}{35}$. Their sum can be obtained simply by adding the two numerators and retaining the common denominator. Thus $\frac{10}{35} + \frac{21}{35} =$ __________ .

\- - - - - - - - - - - - - - -

$\mathbf{\frac{31}{35}}$

An analogous procedure can be used to subtract fractions; for example, in the operation $\frac{4}{5} - \frac{3}{10}$ we can transform $\frac{4}{5}$ into the equivalent fraction __________ so that the two fractions will have a common denominator.

\- - - - - - - - - - - - - - -

$\mathbf{\frac{8}{10}}$

Now we have the difference in the form $\frac{8}{10} - \frac{3}{10}$, which gives the resulting fraction of __________, which can be reduced to the simpler fraction __________.

- - - - - - - - - - - - - - -

$\mathbf{\frac{5}{10}}$; $\mathbf{\frac{1}{2}}$

The same principle applies if letters are involved in the fraction. Suppose that we wish to subtract $\frac{34}{s}$ from $\frac{a}{ns}$. The least common denominator is ns. To get the fraction $\frac{34}{s}$ into a form in which its denominator is ns we multiply both the numerator and denominator of $\frac{34}{s}$ by __________, thus giving the equivalent fraction __________.

- - - - - - - - - - - - - - -

$\boldsymbol{n}$; $\boldsymbol{\frac{34n}{ns}}$

Now our operation is in the form $\frac{a}{ns} - \frac{34n}{ns}$. Since the two fractions have the same denominator, from the first numerator we subtract the second numerator and we keep the common denominator. Therefore $\frac{a}{ns} - \frac{34n}{ns} =$ ________________.

- - - - - - - - - - - - - - -

$$\frac{a - 34n}{ns}$$

We are required to perform the operation $\frac{5k}{2x} + \frac{7}{3y} - \frac{1}{yz}$. The least common denominator is ________________.

- - - - - - - - - - - - - - -

$6xyz$

To get the first fraction of our expression into a form in which its denominator is $6xyz$ we multiply both the numerator and denominator of $\frac{5k}{2x}$ by __________, thus giving the equivalent fraction __________________.

\- - - - - - - - - - - - - - -

$3yz$; $\frac{15kyz}{6xyz}$.

To get the second fraction into a form in which its denominator is $6xyz$ we multiply the numerator and denominator of $\frac{7}{3y}$ each by ____________, giving the equivalent fraction ____________.

\- - - - - - - - - - - - - - -

$2xz$; $\frac{14xz}{6xyz}$

To get the third fraction with a denominator of $6xyz$ we multiply the numerator and denominator of $-\frac{1}{yz}$ each by _________,

giving the equivalent fraction ________________.

- - - - - - - - - - - - - - -

$6x$; $-\frac{6x}{6xyz}$

Assembling the three equivalent fractions, we now have the problem stated in the form, $\frac{15kyz}{6xyz} + \frac{14xz}{6xyz} - \frac{6x}{6xyz}$. Combining the numerators and retaining the common denominator, we have for an

answer the fraction ________________________________.

- - - - - - - - - - - - - - -

$$\frac{15kyz + 14xz - 6x}{6xyz}$$

Notice that no further combining of the three terms in the numerator can be accomplished because each term involves a different combination of letters.

MULTIPLICATION AND DIVISION OF FRACTIONS

We can multiply fractions by simply multiplying their corresponding numerators and denominators; for example, $\frac{3}{8}(\frac{7}{5}) = \frac{21}{40}$.

Thus $\frac{5}{2}(\frac{7}{11})(\frac{1}{3}) =$ _______________.

- - - - - - - - - - - - - - -

$$\frac{35}{66}$$

In the same fashion $\frac{5a}{8}(\frac{13}{9L}) =$ _______________.

- - - - - - - - - - - - - - -

$$\frac{65a}{72L}$$

To say that one-third of 21 is 7 is equivalent to saying that 21 divided by 3 is 7. One-third can be written as the fraction $\frac{1}{3}$, which is the "reciprocal" of 3. The reciprocal of any number is plus one divided by that number. Thus the reciprocal of 18 is the fraction __________.

- - - - - - - - - - - - - -

$$\mathbf{\frac{1}{18}}$$

The reciprocal of d is $\frac{1}{d}$. The reciprocal of $83u^2v$ is

__________________.

- - - - - - - - - - - - - -

$$\mathbf{\frac{1}{83u^2v}}$$

As we said before, 21 divided by 3 is 7. Let us express this operation in the following equivalent form: $\frac{21}{1} \div \frac{3}{1} = 7$. We also said that one-third of 21 is 7, which can be written as $\frac{21}{1}(\frac{1}{3}) = 7$. Notice that we obtained the same numerical result by *multiplying* $\frac{21}{1}$ by the *reciprocal* of 3. Speaking more generally, we can divide two fractions by *inverting* the divisor (the second fraction in sequence) and multiplying; for example, the operation $\frac{5}{7} \div \frac{3}{8}$ can be transformed to $\frac{5}{7}(\frac{8}{3})$ to give the answer $\frac{40}{21}$. If we wish to perform the operation $\frac{4}{17} \div \frac{3}{5}$, we can first invert the divisor, thus transforming this division problem into a problem involving the *product* of two fractions, namely, ________ (________).

- - - - - - - - - - - - - - - -

$\mathbf{\frac{4}{17}; \frac{5}{3}}$

Thus we have the operation $\frac{4}{17}(\frac{5}{3})$, which gives an answer of

____________________.

- - - - - - - - - - - - - - -

$\mathbf{\frac{20}{51}}$

By inverting the divisor, the operation $\frac{6pq}{7} \div \frac{3p}{11}$ can be trans-

formed to the product ____________________________________.

- - - - - - - - - - - - - - -

$\mathbf{\frac{6pq}{7}(\frac{11}{3p})}$

Multiplying numerators and denominators, we obtain the fraction ________________, which can be reduced to the fraction

________________.

- - - - - - - - - - - - - - -

$\frac{66pq}{21p}$; $\frac{22q}{7}$

The operation $\frac{13C}{m} \div \frac{8}{5G}$ can be transformed to the product

________________ to give the answer ________________.

- - - - - - - - - - - - - - -

$\frac{13C}{m}\left(\frac{5G}{8}\right)$; $\frac{65CG}{8m}$

MULTIPLICATION AND DIVISION OF DECIMALS

For many purposes it is convenient to transform fractions into decimals, which imply a denominator of 10 or some power of 10; for example, the fraction $\frac{3}{5}$ is equivalent to $\frac{6}{10}$. This can be written as .6, which is a decimal. If we express the fraction $\frac{13}{25}$ as an equivalent fraction with a denominator of 100, the numerator must be ____________.

- - - - - - - - - - - - - - -

52

$\frac{52}{100}$ can be written in decimal form as ____________.

- - - - - - - - - - - - - - -

.52

The decimal .52 is said to have two "decimal places." The decimal 14.13 also has two decimal places. The product of two decimals will have as many decimal places as the *sum* of the number of decimal places in the factors; for example, .23(.2) gives .046 for an answer. There are two decimal places in .23 and one decimal place in .2; hence the product must have three decimal places. Notice that we had to supply a zero to indicate the extra decimal place: 6.1(.3) = ____________.

- - - - - - - - - - - - - - -

1.83

73.2(.0004) = ____________________.

- - - - - - - - - - - - - - -

.02928

50.03(200) = _____________ and 7.34(1.21) = _____________.

- - - - - - - - - - - - - - -

10,006 [equivalent to **10,006.00**]; **8.8814**

To divide one decimal by another decimal an efficient procecedure is to multiply the divisor by a power of 10 so that it becomes an "integer" (i.e., a "whole number"); for example, if we wish to divide 36 by 1.2, the answer is equivalent to the fraction $\frac{36}{1.2}$, which in turn is equivalent to the fraction $\frac{360}{12}$. This is because we multiplied the numerator and denominator of $\frac{36}{1.2}$ each by__________.

\- - - - - - - - - - - - - - -

10

$\frac{360}{12} =$ __________, so $\frac{36}{1.2}$ also equals__________.

\- - - - - - - - - - - - - - -

30; 30

If we had to divide 62.4 by 3.12, we could multiply each number by 100, giving us the fraction $\frac{6240}{312}$, which equals__________.

\- - - - - - - - - - - - - - -

20

This procedure is particularly useful if the division of numbers requires "long division"; for example, if we are to divide 2.451 by 4.3, we can set it up first as $4.3\overline{)2.451}$ and then as $43\overline{)24.51}$. The answer to this division problem is __________.

- - - - - - - - - - - - - - -

.57

Notice that we moved the decimal point one place to the right in *both* dividend and divisor. Suppose that we are dividing .0962 by .013. By multiplying each number by 1000 we can set this up as $13\overline{)96.2}$, which gives the answer __________.

- - - - - - - - - - - - - - -

7.4

In this latest example we effectively multiplied the dividend and divisor each by 1000 by moving the decimal point __________ places to the right.

- - - - - - - - - - - - - - -

3

If the division does not "come out evenly," we affix as many zeros to the right of the dividend as is necessary in view of the number of figures desired in the quotient; for example, if we wish to divide 2.31 by 7.4, we can set it up as $74\overline{)23.10}$, giving .31, which is accurate to two decimal places; .12 divided by .04 equals ________.

- - - - - - - - - - - - - - -

3

---

750 divided by 1.5 = ________ and 19.034 divided by 3.07 = ________.

- - - - - - - - - - - - - - -

500; 6.2

Carried to one decimal place, 4.57 divided by .081 = ________.

- - - - - - - - - - - - - - -

56.4

BASIC OPERATIONS APPLIED TO SIGNED NUMBERS

Negative numbers can be used to indicate distances that are opposite in direction to those indicated by positive numbers. To add numbers with like algebraic signs we add the quantities and affix the common sign; for example, $7 + 25 + 12 = 44$. Likewise $(-13) + (-8) + (-5) + (-6) = -32$. Thus $(-4) + (-9) + (-11) =$ ____________.

\- - - - - - - - - - - - - - -

−24

To add two numbers with unlike signs, we take the difference between the two quantities and affix the sign of the larger quantity; for example, $7 + (-3) = 4$; $8 + (-15) = -7$; $-24 + 13 = -11$; $-9 + 36 = 27$; $19 + (-33) =$ ____________.

\- - - - - - - - - - - - - - -

−14

$-3.7 + 8.3 = 4.6$; $-0.51 + 3.18 =$ ____________.

\- - - - - - - - - - - - - - -

2.67

Often, when a series of numbers with unlike signs is to be added, we add the positive numbers and negative numbers separately and then combine the partial sums. Consider, for example, the problem, $8 - 12 - 5 + 14 - 9$. First we add the numbers with positive signs; 8 plus 14 gives __________.

\- - - - - - - - - - - - - - -

22

Next we add the numbers with negative signs; 12 plus 5 plus 9 gives __________.

\- - - - - - - - - - - - - - -

26

Now we combine the two partial sums, prefixing each with its appropriate algebraic sign. Thus we have $+22$ -26 and the algebraic sum of all the original numbers is __________.

\- - - - - - - - - - - - - - -

–4

In the problem $-7.3 + 6.1 + 8.4 - 3.9$, the positive partial sum is ___________, and the negative partial sum is 11.2.

- - - - - - - - - - - - - - -

14.5

Now we have the problem in the form $14.5 - 11.2$, so the final answer is ___________.

- - - - - - - - - - - - - - -

3.3

To subtract one signed number from another we change the sign of the "subtrahend" (the number to the right or below) and *add* according to the foregoing rules; for example, $12 - (-7) = 19$. Notice that in a sense the two minus signs "cancel" the effect of each other. Likewise, $4.5 - (-6.1) = 10.6$; $12.5 - (-9.8) =$ ___________.

- - - - - - - - - - - - - - -

22.3

$-38y - (-14y) = -24y$; $-1.51 - (-7.83) = 6.32$; $-8.2L - (-5.9L) =$ _______________.

- - - - - - - - - - - - - - -

$-2.3L$

The multiplication (or division) of numbers with *like* algebraic signs gives a positive product (or quotient). Thus, $13(7) = 91$; $125/25 = 5$; $-4(-13) = 52$; $-2.64/-3.3 =$ __________.

- - - - - - - - - - - - - - -

.8

The multiplication (or division) of numbers with *unlike* algebraic signs gives a negative product (or quotient); for example, $45(-3) = -135$; $-14(11) = -154$; $-8/2 = -4$; $117/-13 =$ __________.

- - - - - - - - - - - - - - -

-9

$$\frac{-49Qt^2u}{56tu^2} = \underline{\hspace{4cm}}.$$

- - - - - - - - - - - - - - -

$$-\frac{7Qt}{8u}$$

At this point take the following review test to find out how well you have learned the preceding material. Do your work on a separate sheet of paper. The answers can be found on p. 159.

FIRST REVIEW TEST

1. $\frac{1}{4} + \frac{1}{3} =$

2. $\frac{1}{6} + \frac{5}{9} =$

3. $\frac{1}{4} - \frac{1}{6} =$

4. $\frac{13}{18} - \frac{5}{12} =$

5. $\frac{3}{x} + \frac{7}{y} =$

6. $\frac{1}{4k} - \frac{1}{2n} =$

7. $\frac{2}{3C} - \frac{3}{4} + \frac{1}{2t} =$

8. $\frac{2}{3}(\frac{4}{5}) =$

9. $\frac{7}{9}(\frac{10}{13}) =$

10. $\frac{1}{B}(\frac{3f}{5})(\frac{2}{7s}) =$

11. $\frac{4}{9} \div \frac{5}{7} =$

12. $\frac{14}{5} \div 3 =$

13. $7M \div \frac{2}{9} =$

14. $\frac{4x}{3} \div \frac{2x}{3} =$

15. $3.8(1.6) =$

16. $5.09(2.7) =$

17. $1.4(.0003) =$

18. $84/1.2 =$

19. $18.225/2.43 =$

20. $.03596/.0062 =$

21. $25 + (-18) =$

22. $-43 + 15 =$

23. $7.6 - 2.7 + 8.3 =$

24. $5 - 12 + 8 - 9 - 24 + 17 + 2 - 15 =$

25. $3.64 - (-5.91) =$

26. $6.24(-3.1) =$

27. $-7.05(-18) =$

28. $57.6/-9 =$

29. $-2.89/-1.7 =$

30. $91ab^2c/-63abd =$

For each problem that you missed check back to the relevant section of the preceding material to find out how you went wrong.

PARENTHESES AND BRACKETS

When more than two numbers are to be operated on algebraically, there can be ambiguity about the order of operations. Therefore parentheses and brackets are used. The operations within parentheses should be performed first; for example, $3(8 + 4) = 3(12) = 36$; $(17 - 9)x = 8x$; $(5 - 13 + 6)4 = (\_\_\_\_)4$ $= \_\_\_\_$.

- - - - - - - - - - - - - - -

–2; –8

$-12(27 + 4 - 8) = \_\_\_\_$.

- - - - - - - - - - - - - - -

–276

In the expression $7(a - 3)$, in which one of the terms within parentheses is an unknown, the factor 7 is multiplied separately by the terms within the parentheses; that is, $7(a - 3) = 7a - 21$; $(z + 1.6)23 = \_\_\_\_$.

- - - - - - - - - - - - - - -

$23z + 36.8$

Sometimes in complex operations brackets are used as well as parentheses. The operations within brackets should be performed *after* the operations within parentheses; for example, $[8(7 - 2)] - 13 = [8(5)] - 13 = 40 - 13 = 27$; $28y[2(6 - 14)] = 28y[2(\_\_\_\_\_\_\_\_)]$.

- - - - - - - - - - - - - - - -

−8

$28y[2(-8)] = 28y[\_\_\_\_\_\_\_\_] = \_\_\_\_\_\_\_\_$.

- - - - - - - - - - - - - - - -

−16; −448*y*

$$7\left[\frac{3n}{4} - (5)^2\right] = 7\left[\frac{3n}{4} - 25\right] = \frac{21n}{4} - 175;\ [5x(x - 11)] - 2x^2$$

$= [\_\_\_\_\_\_\_\_ - \_\_\_\_\_\_\_\_] - 2x^2 = \_\_\_\_\_\_\_\_ - \_\_\_\_\_\_\_\_$.

- - - - - - - - - - - - - - - -

$5x^2$; $55x$; $3x^2$; $55x$

If a minus sign precedes an expression within parentheses, the parentheses can be removed if the signs of all the terms within the parentheses are changed; for example, $9 - (12 - 7) = 9 - 12 + 7 = 4$. Notice that this is the same result that would have occurred if we had combined within parentheses first; that is, $9 - (12 - 7) = 9 - 5 = 4$. This principle is particularly useful when unknowns are contained within parentheses; for example, $13B - (45 + 8B) = 13B - 45 - 8B = 5B - 45$; $28 - (19 - 2m) =$ ______________.

- - - - - - - - - - - - - - -

$9 + 2m$

USE OF ZERO

Adding or subtracting zero from any number leaves the number unchanged; for example, $14 + 0 = 14$ and $7c^2 - 0 = 7c^2$. Multiplying any number by zero gives zero; for example, $0(28) = 0$; $(35F^2 - 4z)0 =$ __________.

- - - - - - - - - - - - - - -

0

Zero divided by any number (except zero itself) gives zero; for example, $\frac{0}{7} = 0$; $\frac{0}{91v} =$ __________.

- - - - - - - - - - - - - - -

0

The use of zero as a divisor is not permitted! In applied mathematical work, if a series of operations seems to result in the necessity to divide by zero, it is likely that some mistake has been made in the previous calculations.

BASIC OPERATIONS APPLIED TO SQUARE ROOT RADICALS

In statistical work squares and square roots are used often. In the equation $\sqrt{49} = 7$ the number under the radical sign is called the "radicand," which in this equation is 49. In the equation $\sqrt{144} = 12$ the radicand is ________.

- - - - - - - - - - - - - - -

144

Although $(-12)^2 = 144$ as well as $(+12)^2 = 144$, in statistical work it is always assumed that the positive square root is indicated. To multiply two radicals write the product of their radicands under the radical sign; for example, $\sqrt{4}\,\sqrt{16} = \sqrt{64} = 8$; $\sqrt{xy}\,\sqrt{xz} = \sqrt{x^2yz}$; $\sqrt{2a}\,\sqrt{7c} =$ ____________________.

- - - - - - - - - - - - - - -

$\sqrt{\mathbf{14}ac}$

To divide two radicals write the quotient of their radicands under the radical sign; for example,

$$\frac{\sqrt{81}}{\sqrt{9}} = \sqrt{\frac{81}{9}} = \sqrt{9} = 3; \quad \frac{\sqrt{63y^2}}{\sqrt{9y}} = \sqrt{7y}; \quad \frac{\sqrt{75b^2}}{\sqrt{3}} = \sqrt{\underline{\qquad\qquad}}.$$

- - - - - - - - - - - - - - -

$25b^2$

Taking the positive square root, we have $\sqrt{25b^2} =$ ________.

- - - - - - - - - - - - - - -

$5b$

It is sometimes useful to move a factor under the square root radical. To do this square the factor, multiply it by the radicand, and then write the product under the radical; for example,

$$2\sqrt{7} = \sqrt{4(7)} = \sqrt{28}; \quad \frac{1}{n}\sqrt{35 + 4n} = \sqrt{\frac{1}{n^2}(35 + 4n)} = \sqrt{\frac{35}{n^2} + \frac{4}{n}};$$

$$2f\sqrt{7f - 8} = \sqrt{\underline{\qquad\qquad}(7f - 8)}.$$

- - - - - - - - - - - - - - -

$4f^2$

$$\sqrt{4f^2(7f-8)} = \sqrt{\_\_\_\_\_\_\_\_\_\_}.$$

- - - - - - - - - - - - - - -

$28f^3 - 32f^2$

$$\frac{5}{N}\sqrt{3N - 10N^2} = \sqrt{\frac{\quad}{\_\_\_\_}(3N - 10N^2)}.$$

- - - - - - - - - - - - - - -

$\frac{25}{N^2}$

$$\sqrt{\frac{25}{N^2}(3N - 10N^2)} = \sqrt{\_\_\_\_\_\_\_\_\_\_}.$$

- - - - - - - - - - - - - - -

$\frac{75}{N} - 250$

The square root of a number between zero and one is always greater than the number itself; for example, in the equation $\sqrt{.25} = .5$ the square root .5 exceeds .25, which is the number itself. Likewise, $\sqrt{.01} = .1$; $\sqrt{.09} =$ __________.

- - - - - - - - - - - - - -

.3

$\sqrt{.0144} =$ ____________.

- - - - - - - - - - - - - -

.12

SOLUTION OF LINEAR EQUATIONS

The usual procedure in solving linear equations in one unknown is to move all terms involving the unknown to the left side of the equation by adding or subtracting and then to remove the coefficients by multiplying or dividing. Consider the equation $7y - 13 = 4y + 5$. We want the terms involving y to be on the left side and the other terms to be on the right side. We can first add 13 to *each* side of the equation to give $(7y - 13) + 13 = (4y + 5) + 13$. Carrying out the addition on each side, we have

__________ = ____________________ .

\- - - - - - - - - - - - - - -

$7y$; $4y + 18$

Notice that this same equation would have resulted if we had transposed the term (-13) in the original equation to the other side of the equation, meanwhile changing its sign:

$$7y - 13 = 4y + 5$$
$$7y = 4y + 5 + 13$$
$$7y = 4y + \_\_\_\_\_\_ .$$

\- - - - - - - - - - - - - - -

18

In fact, persons who solve many such equations are more likely to think in this fashion. It should be remembered, however, that the rationale for this procedure of transposition is the process of adding to or subtracting from both sides of the equation.

Continuing with our example, we transpose the $4y$ term to the left side, meanwhile changing its sign:

$$7y = 4y + 18.$$
$$7y - 4y = 18$$
$$\underline{\qquad\quad} = 18$$

- - - - - - - - - - - - - - -

3y

We now have the equation $3y = 18$. To get y by itself on the left side of the equation we can divide *both* sides of the equation by 3. This gives the equation $y =$ __________, which is our final result.

- - - - - - - - - - - - - - -

6

This example illustrates the general rule for solving equations; namely, whichever operation is performed on one side of an equation must also be performed on the other.

As another example, consider the equation $\frac{z}{3} + 5 = \frac{3}{4}$. Although there are alternative procedures in solving equations involving fractions, it is usually more efficient to "clear all fractions" at the outset. In the present situation we can first multiply both sides of the equation by 12. This requires that *each term* on both sides of the equation be multiplied by 12. Multiplying $\frac{z}{3}$ by 12 gives us __________.

\- - - - - - - - - - - - - - -

4*z*

Likewise, multiplying 5 by 12 gives __________, and multiplying $\frac{3}{4}$ by 12 gives __________.

\- - - - - - - - - - - - - - -

60; 9

Thus the effect of multiplying the entire equation $\frac{z}{3} + 5 = \frac{3}{4}$ by 12 is to produce the new equation __________ + __________ = __________.

\- - - - - - - - - - - - - - -

$4z$; 60; 9

We now have the equation $4z + 60 = 9$. This can be transformed to $4z =$ __________.

\- - - - - - - - - - - - - - -

-51

Having the equation $4z = -51$, we need only divide by 4, which produces the final result $z =$ ____________________.

\- - - - - - - - - - - - - - -

-12.75 [or] **$-12\frac{3}{4}$**

Suppose that we wish to solve for x in the equation $6ax = 12a^2 - 7$. Dividing the equation by 6a gives $x = \frac{12a^2 - 7}{6a}$. The right side of the equation can be transformed by dividing *each term* in the numerator by 6a, thus giving x = ______ – ______.

- - - - - - - - - - - - - - -

2a; $\mathbf{\frac{7}{6a}}$

The important point here is that *all* terms in the numerator must be divided by the single term in the denominator. It is generally impossible to simplify fractions with one term in the numerator and more than one term in the denominator; for example, $\frac{3}{2 - x}$.

To solve the equation $\sqrt{x} = a - b$ we square both sides of the equation to give $x = a^2 - 2ab + b^2$. Notice carefully the "cross-product term" $-2ab$. Given the equation, $\sqrt{v} = 3 + 5m$, v = ______________________.

- - - - - - - - - - - - - - -

$\mathbf{9 + 30m + 25m^2}$

To solve $\frac{y^2}{3} = 12$ we can first multiply by 3 to give $y^2 =$ __________ and $y =$ __________.

- - - - - - - - - - - - - - -

36; 6 [The negative square root is not used in statistics.]

In general, however, the square root of the sum or difference of terms *cannot* be simplified; for example,

$$\sqrt{16 + \frac{9}{b^2}} \text{ does } NOT \text{ equal } 4 + \frac{3}{b}!$$

Now let us solve an equation that involves several of the foregoing principles. This equation is actually used in applied statistics, although different symbols are involved. The material is more difficult than the preceding material. If the reader is unable to master it, he should not be greatly concerned because he will probably still be able to do the problems in the Review Test with facility.

Given the equation $a^2 (1 - x^2) = b^2 (1 - c^2)$, solve for x. We wish to move all terms except those involving x to the right side of the equation. On the left side we have a term involving x multiplied by a factor that does *not* involve x. The strategy calls for disentangling the factor a^2 first. We can do this by __________________ both sides of the equation by a^2.

- - - - - - - - - - - - - - -

dividing

Dividing both sides of the equation $a^2(1 - x^2) = b^2(1 - c^2)$

by a^2 gives ________________ = ________________________.

- - - - - - - - - - - - - - -

$1 - x^2$; $\dfrac{b^2(1 - c^2)}{a^2}$

Now we have $1 - x^2 = \dfrac{b^2(1 - c^2)}{a^2}$. Subtracting 1 from each side gives $-x^2 = \dfrac{b^2(1 - c^2)}{a^2} - 1$. We can now multiply each side by____________.

- - - - - - - - - - - - - - -

-1

Multiplying by −1 gives $x^2 = 1 - \dfrac{b^2(1 - c^2)}{a^2}$. Now we wish to express all terms on the right side over a common denominator. We can replace 1 with its equivalent a^2/a^2. This gives $x^2 = \dfrac{a^2}{a^2} - \dfrac{b^2(1 - c^2)}{a^2}$. By combining the two terms on the right side over the common denominator we have

$x^2 =$ ____________________.

- - - - - - - - - - - - - - -

$$\mathbf{\frac{a^2 - b^2(1 - c^2)}{a^2}} \quad [\text{or}] \quad \mathbf{\frac{a^2 - b^2 + b^2c^2}{a^2}}$$

To solve for x we take the square root of both sides of the equation to obtain $x =$ ____________________.

- - - - - - - - - - - - - - -

$$\mathbf{\sqrt{\frac{a^2 - b^2 + b^2c^2}{a^2}}}$$

If desired, we can further simplify the answer algebraically by expressing it as a product of two square root radicals, $\sqrt{\frac{1}{a^2}}$ $\sqrt{a^2 - b^2 + b^2c^2}$. Taking the square root of the first radical, we obtain the answer $x =$ ______________________.

- - - - - - - - - - - - - - -

$\frac{1}{a}\sqrt{a^2 - b^2 + b^2c^2}$

SOME PROBLEMS ON LINEAR EQUATIONS

Now you have all the "equipment" necessary to solve linear equations involving fractions, decimals, signed numbers, and square root radicals. To see how well you have learned this material try the five problems below. The first three are relatively simple, and the last two are more difficult.

Given $5u - 3 = 17$, $u =$ __________.

- - - - - - - - - - - - - - -

4

Given $7z + 12 = 8$, $z =$ __________.

- - - - - - - - - - - - - - -

−.57 [or] $-\frac{4}{7}$

Given $3x - 5 = 7 - 6c$, $x =$ __________.

- - - - - - - - - - - - - - -

$4 - 2c$

Given $2\sqrt{y} = 5.4$, $y =$ ____________.

- - - - - - - - - - - - - - -

7.29

This problem may have caused you some trouble, so let us do it step-by-step. First we square both sides of the equation, thus obtaining

$$4y = 29.16.$$

Now we divide both sides of the equation by 4, giving the final answer,

$$y = 7.29.$$

Try this problem. Given $8 + \dfrac{3v^2}{5} = 5.4a^2 + 10.4$,

$v =$ ________________________.

- - - - - - - - - - - - - - -

$\sqrt{\mathbf{9a^2 + 4}}$

One way to start this problem is to "clear fractions" by multiplying both sides of the equation by 5 to give

$$40 + 3v^2 = 27a^2 + 52$$

Transposing 40 to the right side, we have

$$3v^2 = 27a^2 + 12$$

Dividing by 3, we obtain

$$v^2 = 9a^2 + 4.$$

Taking the square root, we have

$$v = \sqrt{9a^2 + 4}$$

Notice that this answer *cannot* be further simplified because $(3a + 2)^2 = 9a^2 + 12a + 4$. Refer back to page 47, where this point is discussed.

SUMMARY

The numerator and denominator of a fraction can each be *multiplied* by the same number without changing the value of the fraction. Similarly, the numerator and denominator can each be *divided* by the same number. However, *adding* or *subtracting* the same number from the numerator and denominator will in general change the value of a fraction.

Fractions cannot be *added* or *subtracted* unless they have a common denominator. To *multiply* fractions multiply their numerators and multiply their denominators. The "reciprocal" of a number n is the number 1 divided by n. To *divide* a fraction by a second fraction multiply the first fraction by the reciprocal of the second fraction; that is, invert the divisor and multiply.

To multiply numbers involving decimals, point off in the product the sum of the number of decimal places in all of the factors.

To combine numbers with *like* algebraic signs add the quantities and affix the common sign. To combine numbers with *unlike* signs add the numbers with positive signs; then add the numbers with negative signs. Take the difference between these two partial sums and affix the sign of the larger of them. The multiplication (or division) of numbers with *like* signs gives a positive product (or quotient). The multiplication (or division) of numbers with *unlike* signs gives a negative product (or quotient). If a minus sign precedes an expression within parentheses, the parentheses can be removed if the signs of all the numbers within parentheses are reversed.

Adding or subtracting zero from any number leaves the number unchanged. Multiplying any number by zero gives zero. Zero divided by any number gives zero. The use of zero as a divisor is not permitted.

In applied statistics only the positive square root is used. To multiply radicals write under the radical sign the product of their radicands. To multiply (or divide) a square root radical by a number, write the product (quotient) of the radicand and the *square* of the number under the radical sign.

The square root of a number between zero and one is always greater than the number itself.

In performing operations on equations whatever is done to one side of the equation must also be done to the other. A term may be transposed from one side of an equation to the other if its algebraic sign is reversed.

SECOND REVIEW TEST

1. $5(12 - 4) =$
2. $(9 + 2)y =$
3. $(L - 6 + Q)3 =$
4. $4[(9 - 6)5] =$
5. $[7p(8 - 3x)] - 13p =$
6. $12[\frac{7y}{4} - (6 + 5y - 8z)] =$
7. $48 - 5e(0) =$
8. $0/792M^2 =$
9. $\sqrt{5w}\ \sqrt{11}$
10. $\sqrt{42}/\sqrt{6} =$
11. $\sqrt{75Ab^2}/\sqrt{5b} =$

Transform the next three expressions so that each is under a single square root radical.

12. $3x\sqrt{6 - 5x} =$
13. $6\sqrt{\frac{7c}{9} + \frac{5}{12}} =$
14. $\frac{1}{N}\sqrt{NS - T^2} =$
15. $\sqrt{.64} =$
16. $\sqrt{.0225} =$
17. Given $3x + 5 = 20$, $x =$
18. Given $12 + 7y = 15$, $y =$
19. Given $4.9 - 6z = 46.9$, $z =$
20. Given $\frac{12A}{5} + \frac{1}{3} = \frac{5}{6}$, $A =$
21. Given $4w + 7 = 24c - 5$, $w =$
22. Given $3bv = 27b - 14d$, $v =$
23. Given $2\sqrt{n} = 4 - 3a$, $n =$
24. Given $3r = \sqrt{25 + 8r^2}$, $r =$
25. Given $\frac{X - M}{S} = \frac{Y - N}{T}$, $X =$

Chapter II. Plotting Points and Linear Equations on Graphs

OBJECTIVES

Upon completion of this chapter, the reader should be able to do the following:

1. Use correctly the terminology referring to two-dimensional graphs.

2. Plot points on a two-dimensional graph.

3. Plot a linear equation on a two-dimensional graph.

4. Determine the y intercept and slope of a linear equation.

PLOTTING POINTS

In scientific work one of the main activities is study of the relationship between variables. It is often helpful to portray a relationship between two variables by means of a graph. First we draw two "axes" at right angles to each other. The horizontal axis is called the "x axis" and the vertical axis is called the "y axis." Write the appropriate labels on the axes below.

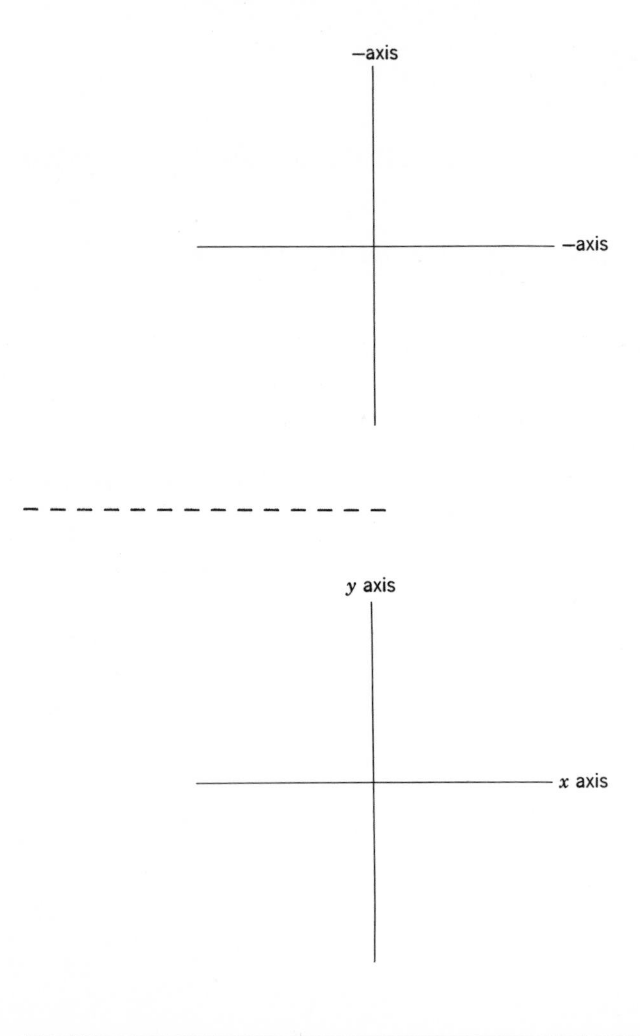

The variable that is measured on the x axis is often called the "independent variable" and the variable that is measured on the y axis is called the "dependent variable." Thus along the horizontal axis we measure the ______________________ variable.

- - - - - - - - - - - - - - -

independent

Along the vertical axis we measure the ______________ variable.

- - - - - - - - - - - - - - -

dependent

The point at which the two axes intersect is called the "origin." Draw an arrow pointing toward the origin on the graph below.

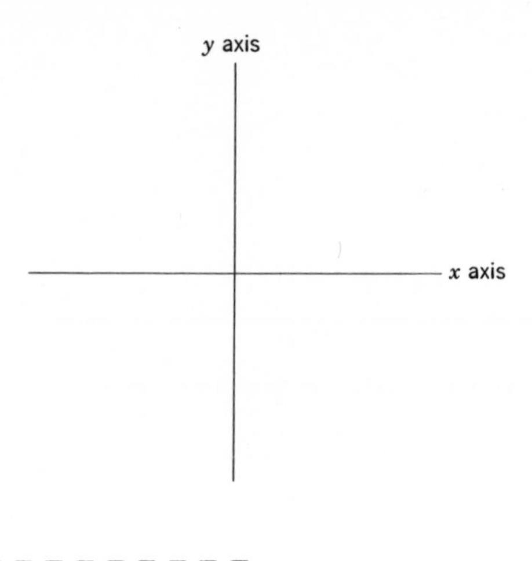

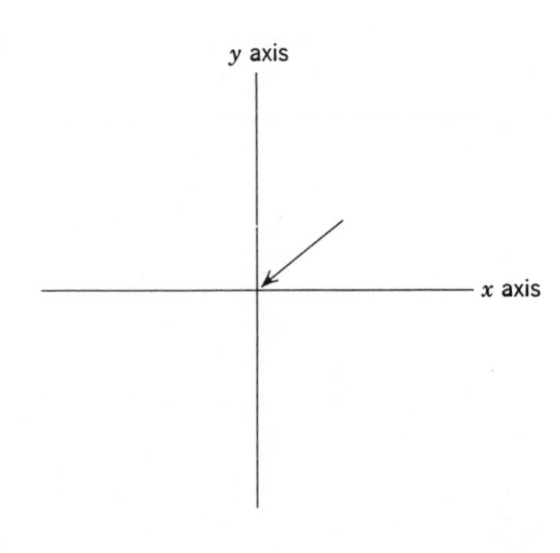

To aid in plotting points on a graph we can "calibrate" the two axes, starting from the origin and proceeding in both directions, as is shown on the graph below.

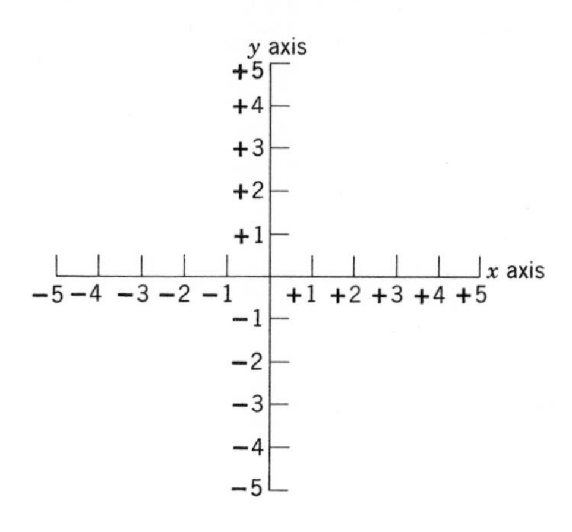

Notice that the positive direction along the x axis is assumed to be toward the *right* and the negative direction along the x axis is assumed to be toward the ________________.

− − − − − − − − − − − − − −

left

Similarly the positive direction along the y axis is assumed to be ________________, whereas the negative direction is assumed to be ____________________.

− − − − − − − − − − − − − −

upward; downward

The location of any point on the graph can be indicated by its two "coordinates." The convention is to represent the x coordinate first within parentheses and the y coordinate second; for example, consider a point with the coordinates (+4, +3). The "x coordinate" for this point is +4 and the "y coordinate" is +3. If a point has coordinates (−2, +3), its x coordinate is ________ and its y coordinate is ________.

- - - - - - - - - - - - - - -

−2; +**3**

Let us first plot the point with the coordinates (+4, +3). Starting from the origin, we move four units to the right along the x axis and then move three units upward and draw a dot. This is shown on the graph below.

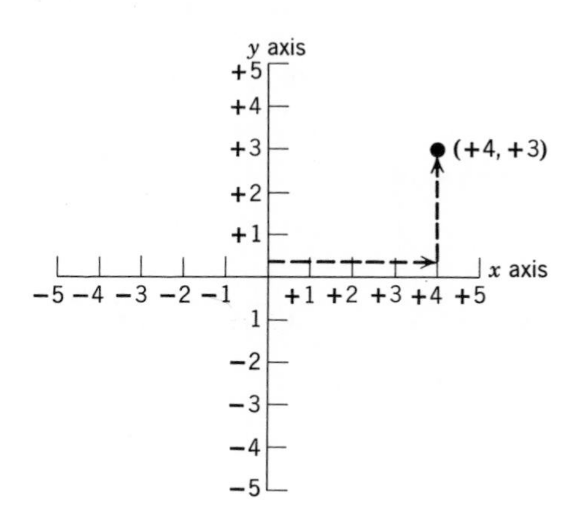

The point plotted above has an x coordinate of _______ and a y coordinate of ______.

- - - - - - - - - - - - - - -

+**4**; +**3**

Now let us plot the point with the coordinates (–2, +3). Again starting from the origin, we move two units to the *left* along the x axis, and three units *upward* and draw a dot. This is shown on the graph below.

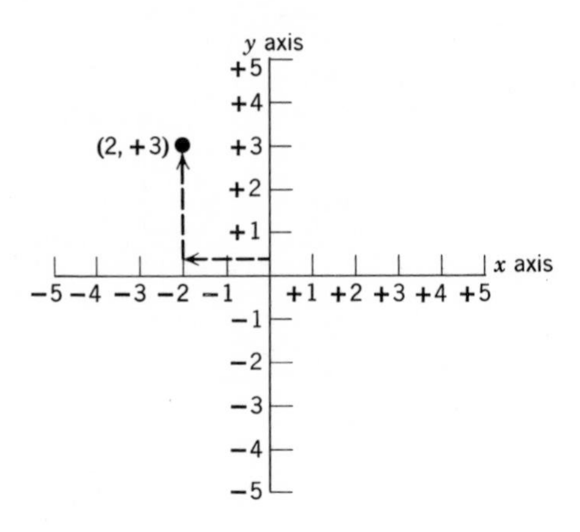

The point plotted above has an x coordinate of ______ and a y coordinate of ______.

- - - - - - - - - - - - - - -

–2; +**3**

Plot the point with the coordinates (+1, −4) on the graph below.

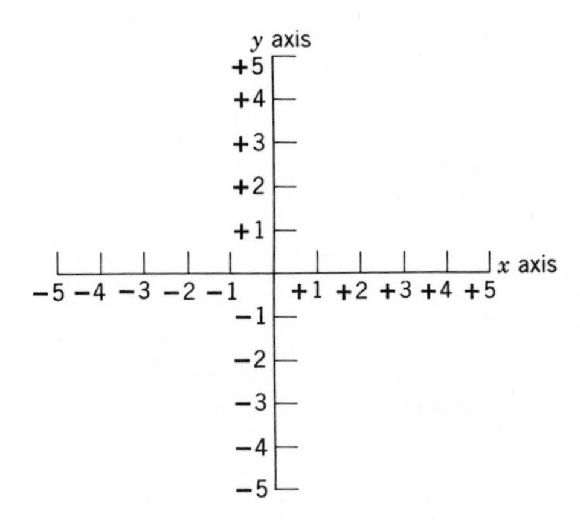

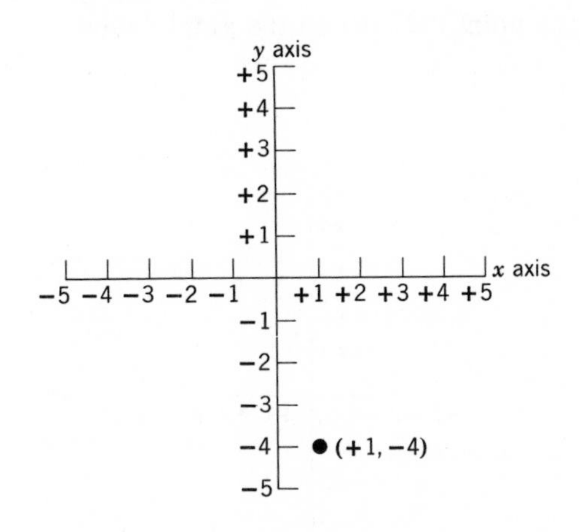

The two axes partition the graph into four quadrants as shown below.

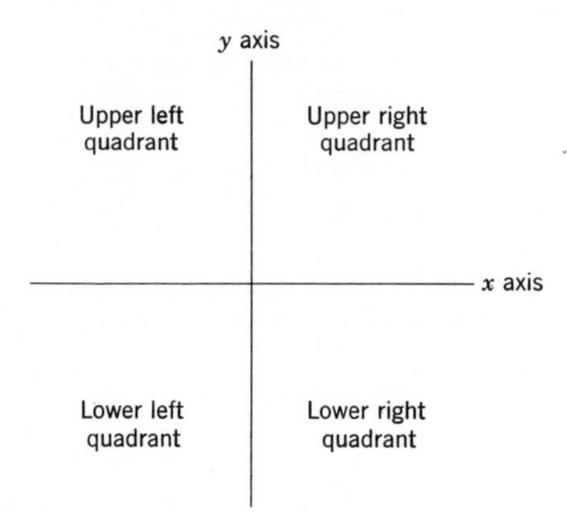

Notice that the point with coordinates (+4, +3) was plotted in the upper right quadrant. Any point, both of whose coordinates are positive, will be located in the upper right quadrant; for example, plot the point (+2, +2) on the graph below.

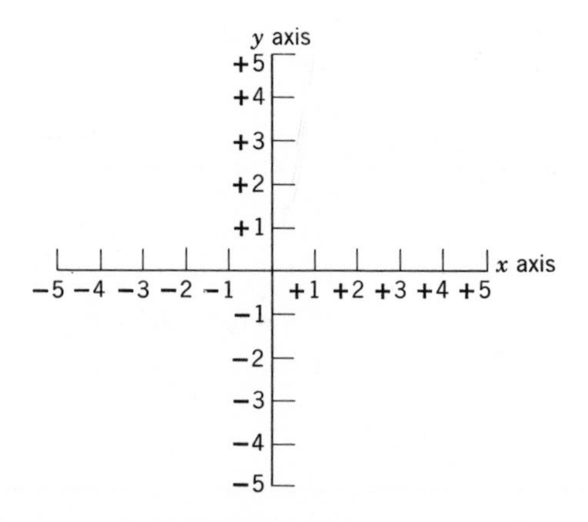

- - - - - - - - - - - - - - -

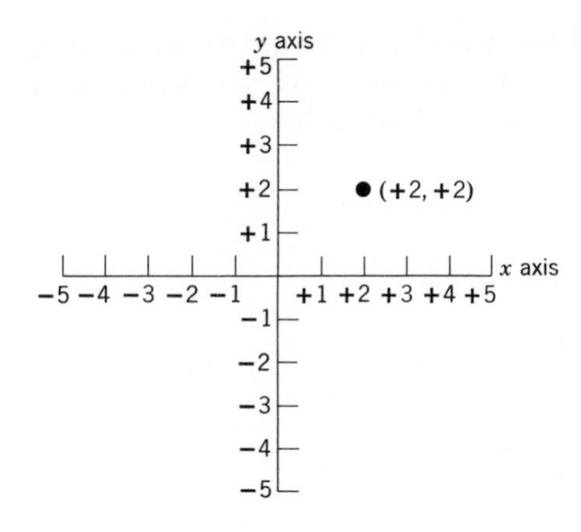

Notice that this point is located in the upper right quadrant. We have not yet plotted any point, *both* of whose coordinates are *negative*, but you can probably figure out from the principles discussed above in which quadrant it will be located. This point will appear in the ______________________ quadrant.

- - - - - - - - - - - - - - -

lower left

There are four possible pairs of the two algebraic signs. There are also four quadrants, and each pair is associated with a particular quadrant as shown below.

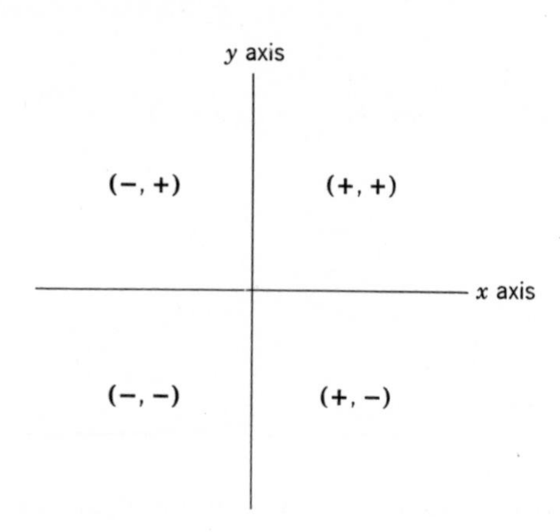

The point with coordinates (+5, −3) would appear in the ______________________ quadrant.

\- - - - - - - - - - - - - -

lower right

The point with coordinates (−1, +8) would appear in the ______________________ quadrant.

\- - - - - - - - - - - - - -

upper left

In practice the positive algebraic sign is usually omitted. Thus the coordinates (7, −3) are equivalent to the coordinates (+7, −3). The coordinates (−2, 9) are equivalent to the coordinates (____________).

\- - - - - - - - - - - - - -

−2, +9

At this stage you may be wondering where the point is plotted if one of the coordinates is zero. Let us plot the point with coordinates (3, 0). Following the principle given above, we should start at the origin, move three units to the right and then zero units upward. This procedure gives the point shown below.

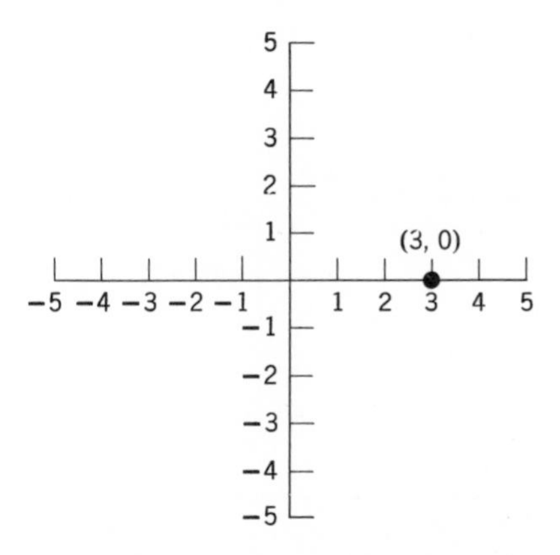

Notice that the point does not lie in any of the quadrants but directly on top of one of the axes. The point with coordinates (3, 0) lies on the x axis. In fact, any point with a y coordinate of zero lies on the x axis. Plot the point with coordinates (0, −2) on the graph on the next page.

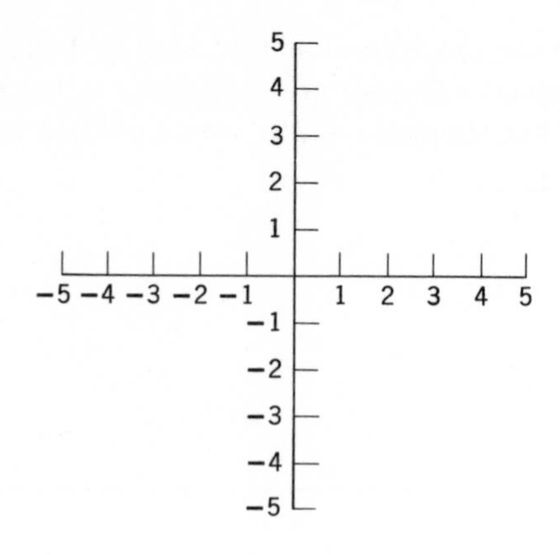
5
4
3
2
1
−5 −4 −3 −2 −1
1 2 3 4 5
−1
−2
−3
−4
−5

– – – – – – – – – – – – – –

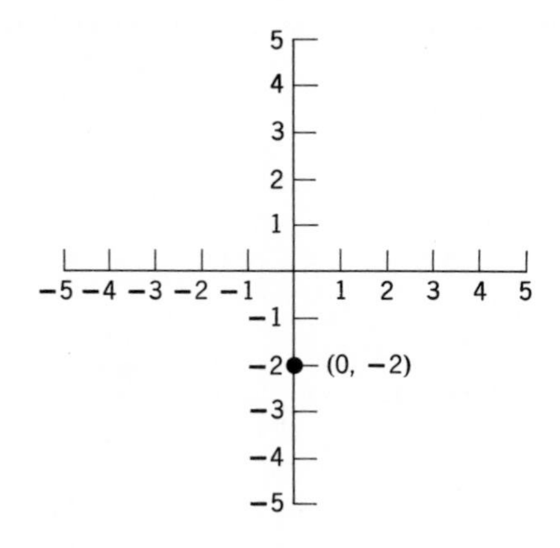
5
4
3
2
1
−5 −4 −3 −2 −1
1 2 3 4 5
−1
−2
(0, −2)
−3
−4
−5

Notice that this point lies on the *y* axis. In fact, any point with an x coordinate of zero lies on the ________ axis.

- - - - - - - - - - - - - -

y

The point (–12, 7) will appear in the ____________________ quadrant.

- - - - - - - - - - - - - -

upper left

The point (–7, –3) will appear in the ____________________ quadrant.

- - - - - - - - - - - - - -

lower left

The point $(0, 5)$ will appear on the ________ axis.

- - - - - - - - - - - - - -

y

The point $(1, 1)$ will appear in the ______________________ quadrant.

- - - - - - - - - - - - - -

upper right

Plot the point (3, –4) on the graph below.

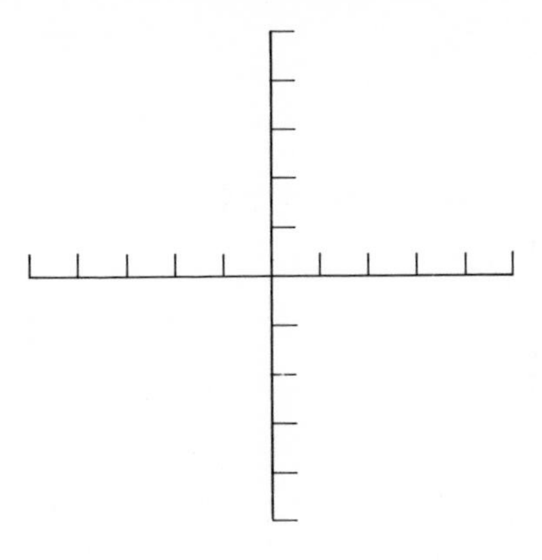

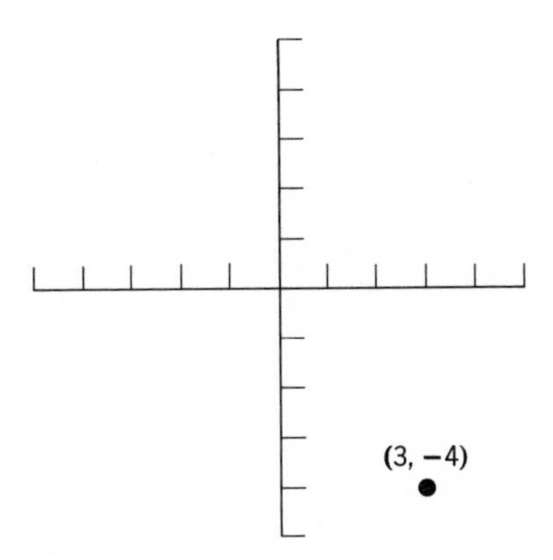

Plot the point (–2, 0) on the graph below.

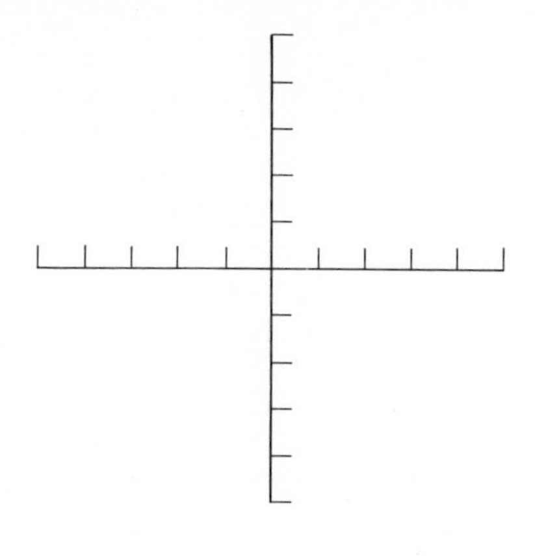

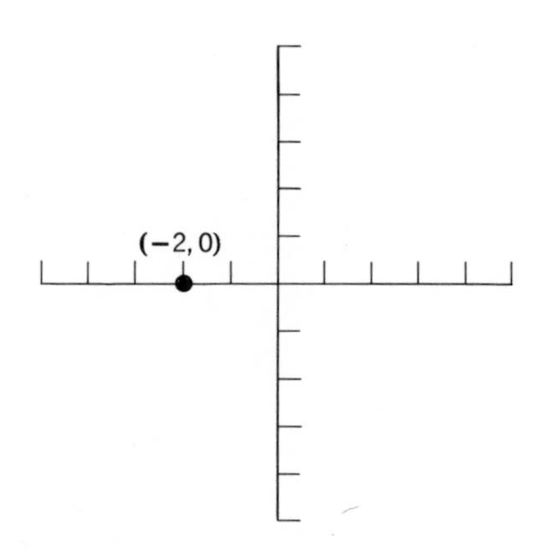

Plot the point $(0, 0)$ on the graph below.

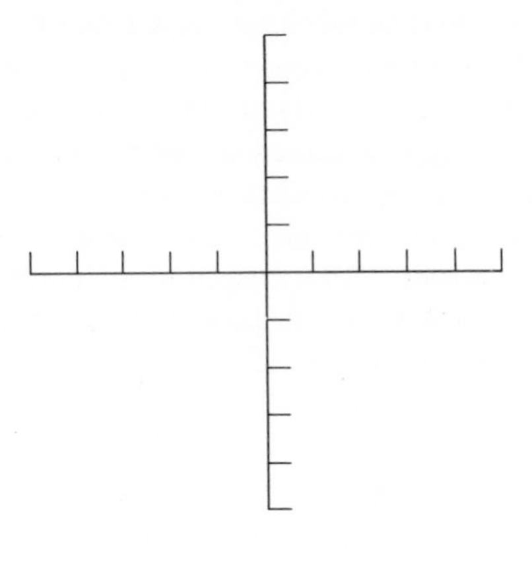

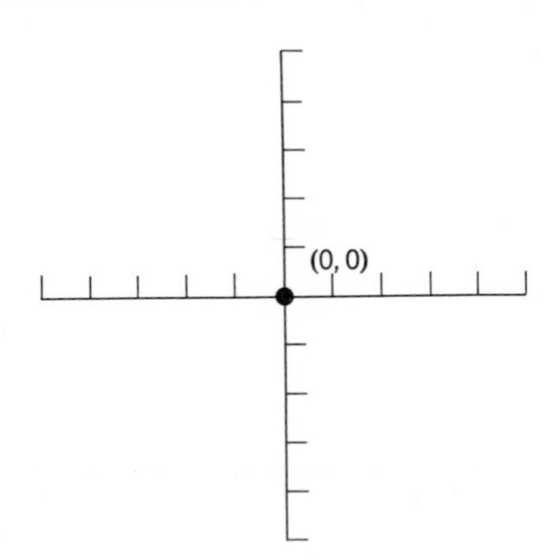

Notice that the point, both of whose coordinates are zero, is plotted directly on top of the origin.

PLOTTING LINEAR EQUATIONS

Graphs can be used to represent equations as well as points. We now consider the graphic representation of a linear equation involving two variables. A "linear equation" is an equation in which the variables appear raised to not more than the first power. An example of a linear equation is $y = 2x + 3$. Since neither x nor y is written with an exponent, it is assumed that they are raised to the first power. An example of a "nonlinear equation" is $y = 5x^2 - 2$. Notice that x is raised to the second power in this equation. The equation $y = 3x^3 + 7$ is a ______________ equation.

\- - - - - - - - - - - - - -

nonlinear

The equation $y = 17x$ is a ______________ equation.

\- - - - - - - - - - - - - -

linear

Let us consider how to represent the equation $y = 2x + 3$ on a graph. First we substitute various values of x in the equation and solve for y; for example, we can substitute 1 for x:

$$y = 2(1) + 3 = 2 + 3 = 5.$$

Thus we have the pair of values $y = 5$ when $x = 1$. This pair of values can be regarded as the coordinates of a point on a two-dimensional graph. Plot the point (1, 5) on the graph below.

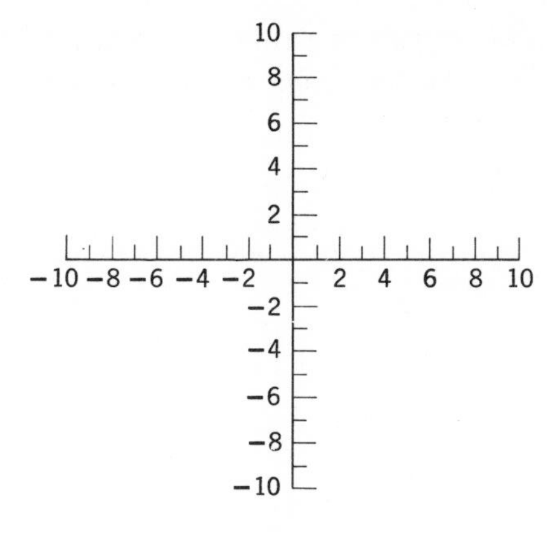

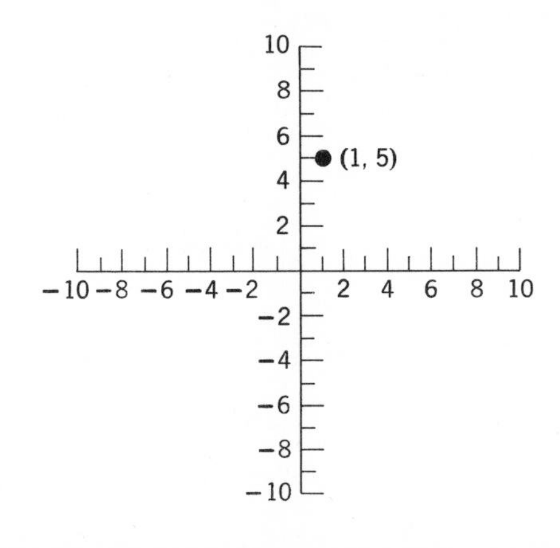

If we substitute 2 for x in the equation, $y = 2x + 3$, we obtain $y =$ ____________.

- - - - - - - - - - - - - - -

7

This gives another pair of values and thus a point with the coordinates (2, 7). Plot this point on the graph below.

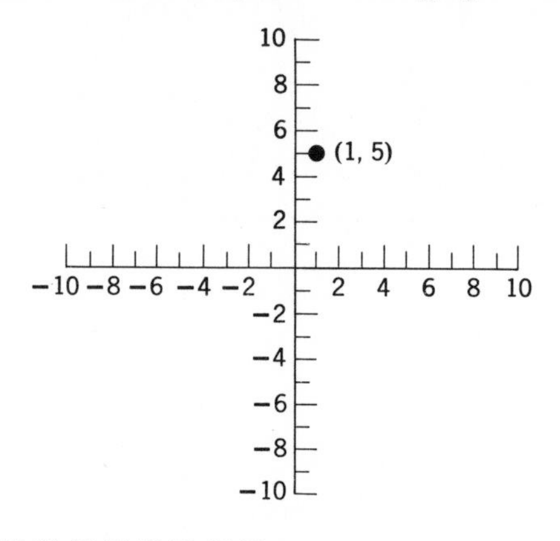

- - - - - - - - - - - - - - -

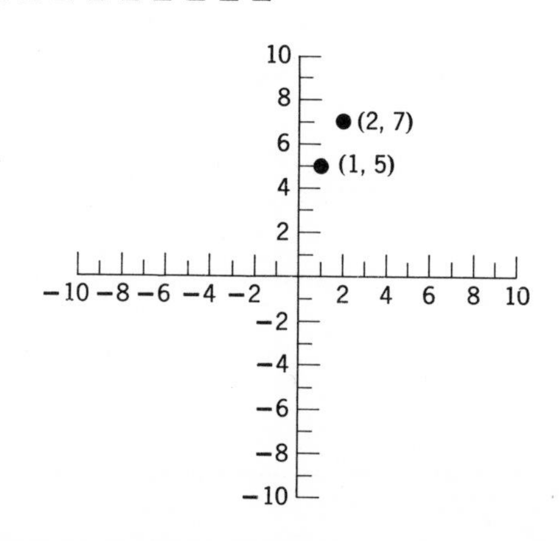

Next substitute $x = 0$ in the equation $y = 2x + 3$. When $x = 0$, $y =$ __________.

- - - - - - - - - - - - - - -

3

Plot the point $(0, 3)$ on the graph below.

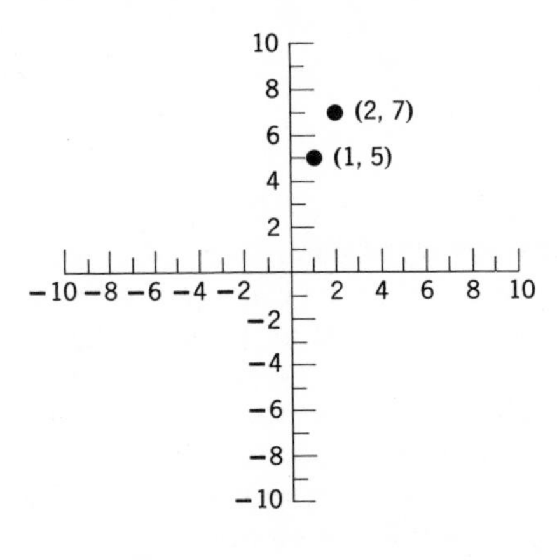

- - - - - - - - - - - - - - -

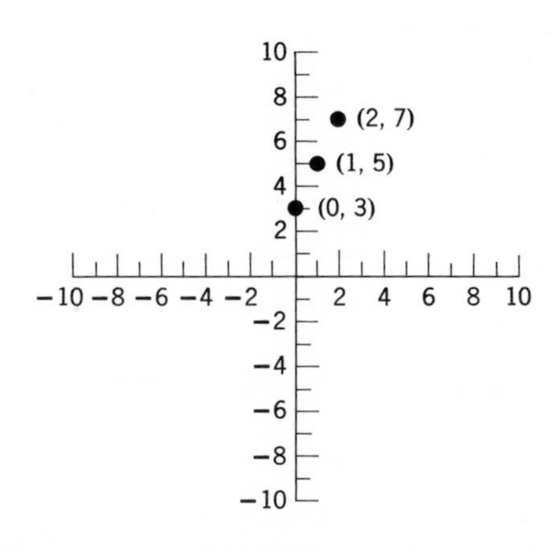

Notice that the three points can be connected by a straight line.

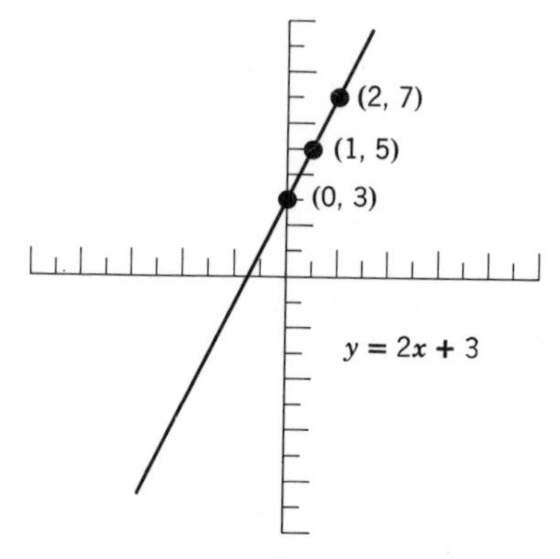

This explains why equations like $y = 2x + 3$ are called "linear equations." Notice that the line crosses the y axis at the point at which $y = 3$. The point at which the line of the equation crosses the y axis cuts off the "y intercept." In this equation the y intercept is 3. Notice that the y intercept appears as a term in the equation of the line, $y = 2x + 3$. It is the number that appears by itself; that is, it is not multiplied by either x or y. In the equation $y = 5x + 4$ the y intercept is __________.

- - - - - - - - - - - - - - -

4

In the equation $y = 3x - 6$ the y intercept is ____________.

- - - - - - - - - - - - - -

−6

Let us now draw the equation line for $y = \frac{1}{2}x + 3$. We shall substitute in turn the values of 0, 2, and 4 for x and solve for y in each case. Substituting $x = 0$, we obtain $y =$ __________.

- - - - - - - - - - - - - -

3

Substituting $x = 2$, we obtain $y =$ __________.

- - - - - - - - - - - - - -

4

Substituting $x = 4$, we obtain $y =$ __________.

- - - - - - - - - - - - - -

5

Now we have three points with coordinates (0, 3), (2, 4), and (4, 5). Plot these points on the graph below and then connect them with a straight line.

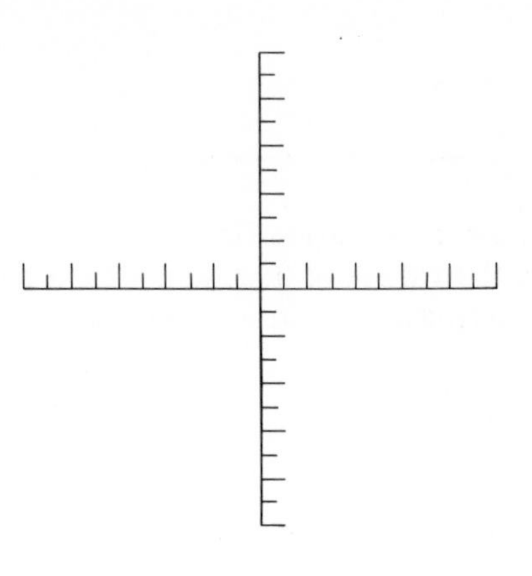

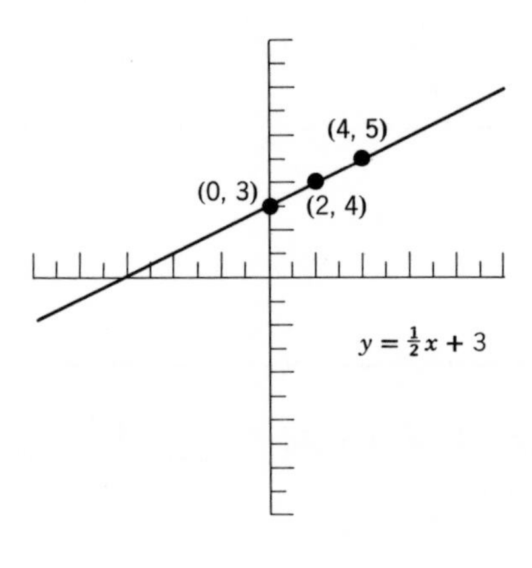

Notice that this equation line also crosses the y axis at the point (0, 3). This is not surprising because inspection of this equation, $y = \frac{1}{2}x + 3$, leads us to conclude that it, as well as the equation $y = 2x + 3$, has a y intercept of 3.

However, notice further that the "slope" of the line for the equation $y = \frac{1}{2} + 3$ is not so steep as the slope of the line for the equation $y = 2x + 3$. In fact, the "coefficient" of x gives an index of steepness for the equation line. The slope of the equation $y = 2x + 3$ is 2. The slope of the equation $y = \frac{1}{2}x + 3$ is

________________.

- - - - - - - - - - - - - - - -

$\frac{1}{2}$

The slope of the equation $y = 3x - 4$ is __________ and its y intercept is ______________.

- - - - - - - - - - - - - - - -

3; −4

Let us now graph the equation $y = x - 2$. Substitute for x the values −1, 0, and 3. We thus obtain for y the corresponding values of __________, __________, and __________.

- - - - - - - - - - - - - - - -

−3; −2; 1

On the graph below plot the points (–1, –3), (0, –2), and (3, 1); then connect them with a straight line.

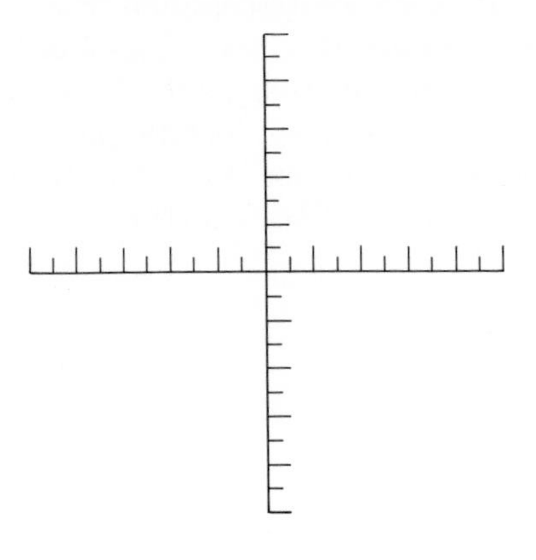

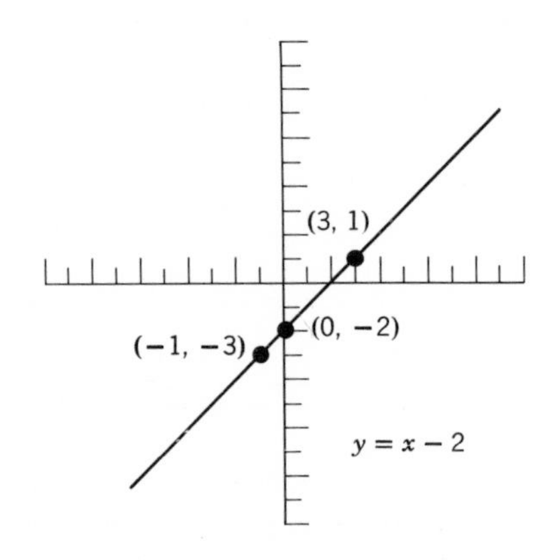

Notice that the line crosses the y axis *below* the origin. We should expect this because inspection of the equation $y = x - 2$ shows the y intercept to be a negative number, namely -2. According to the rule given above, the slope of this equation is

__________.

- - - - - - - - - - - - - - -

1

Comparison of the equation line for the three equations we have drawn so far will show that the slope of this equation is steeper than that of $y = \frac{1}{2}x + 3$ but not so steep as that of $y = 2x + 3$. This is not surprising because the number 1 falls between $\frac{1}{2}$ and 2.

Now we shall graph the equation $y = -2x + 4$. Use these values of x: -1, 0, and 1. The three corresponding points are $(-1,$ __________$)$, $(0,$ __________$)$, and $(1,$ __________$)$.

- - - - - - - - - - - - - - -

6; 4; 2

Draw these points and the line below.

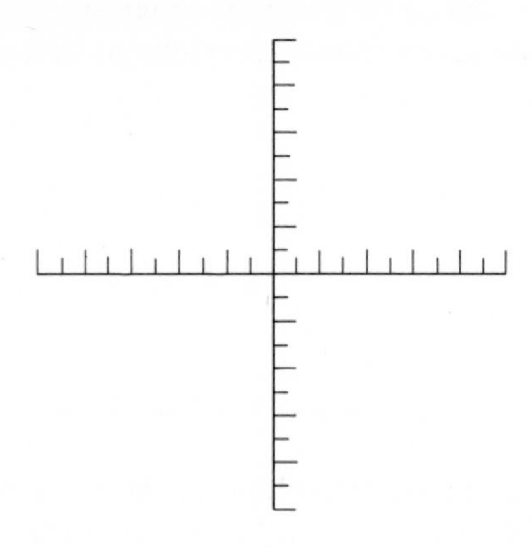

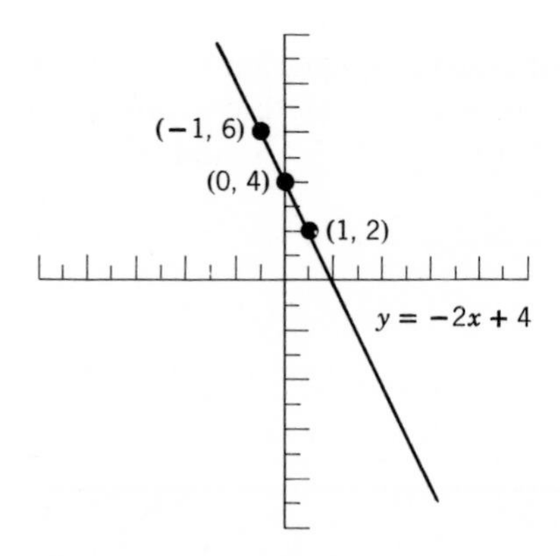

The y intercept is 4, as we note from inspection of the equation. However, the line is tilted in the general direction from upper left to lower right. All of the equation lines we have considered up to now have been tilted from lower left to upper right. This new tilt appears because the equation $y = -2x + 4$ has a *negative* slope, namely −2.

The slope of the equation $y = -\frac{2}{3}x + 6$ is ________.

- - - - - - - - - - - - - - -

$-\frac{2}{3}$

The tilt of this equation would also be in the general direction from upper left to lower right. For the equation $y = 12x - 37$ the slope is ________ and the y intercept is ________.

- - - - - - - - - - - - - - -

12; −37

For simplicity of exposition the equations considered so far have all been presented in a standard form, the form in which the variable y appears by itself, raised to the first power, and on the left side of the equation. In practice you may be presented with linear equations that are not in this standard form. However, by the procedures described in the review of algebra, such equations can be transformed into this form; for example, consider the equation $7x + 3y = 12$. We wish to manipulate this equation until y appears by itself on the left side. First we can transpose $7x$ to the right side to give ______________________.

\- - - - - - - - - - - - - - - -

$\mathbf{3y = -7x + 12}$

Now we can divide both sides by 3 to give

______________________________.

\- - - - - - - - - - - - - - - -

$\mathbf{y = -\frac{7}{3}x + 4}$

The slope of this equation is ____________ and the y intercept is __________.

\- - - - - - - - - - - -

$-\frac{7}{3}$; **4**

Actually, since only two points are needed to define a straight line uniquely, the appropriate equation line can be drawn directly by use of the slope and y intercept without substituting any values of x. The relevant procedures, however, are beyond the scope of the present review.

REVIEW TEST

Using graph paper provided at back of book, plot the points with these coordinates:

1. (2, 3)
2. (6, 4)
3. (0, 5)
4. (−1, 4)
5. (−7, 0)
6. (−5, −4)
7. (0, 0)
8. (−8, 1)

Using the graph paper provided at back of book, plot the lines for these equations:

9. $y = 2x + 1$
10. $y = \frac{3}{4}x + 3$
11. $y = 4x - 2$
12. $3y = 5x + 6$
13. $x = 5 - y$
14. $5x + 2y = -7$

In Problems 9–14 indicate for each problem (a) the y intercept and (b) the slope.

Chapter III. Extraction of the Square Root

OBJECTIVE

Upon completion of this chapter, the reader should be able to extract the square root of a number, accurate to a designated number of significant figures.

EXTRACTION OF THE SQUARE ROOT

The number 7 multiplied by itself gives 49. Forty-nine is said to be the "square" of 7. Seven is said to be the "square root" of 49. Strictly speaking, −7 is also a square root of 49, but in statistics we are concerned only with the positive square root. The square root of 16 is ________.

- - - - - - - - - - - - - - -

4

Thus the square root of a number is another number which, when multiplied by itself, gives the first number. The number 70 squared, which is represented as $(70)^2$, equals ________.

- - - - - - - - - - - - - - -

4900

Calculate the following squares: $(700)^2 =$ ________; $(0.7)^2 =$ ________; $(0.07)^2 =$ ________.

- - - - - - - - - - - - - - -

490,000; 0.49; 0.0049

Now let us arrange these five squares systematically in a table.

$$(700.00)^2 = 49\ 00\ 00.00\ 00$$
$$(70.00)^2 = 49\ 00.00\ 00$$
$$(7.00)^2 = 49.00\ 00$$
$$(.70)^2 = .49\ 00$$
$$(.07)^2 = .00\ 49$$

The extra zeros and spaces have been introduced to illustrate a principle. Notice that for every move of the decimal point in the number the decimal point moves *two* places in the same direction in the square of the number. This is the reason why the squares above have been arranged in "periods" of two digits apiece. Notice also that the same nonzero digits, 4 and 9, recur in the series of squares and that the same digit, 7, recurs in the series of numbers that constitutes their square roots.

Suppose, however, that we need the square root of the number 490. Its square root, rounded to three significant figures, is 22.1. The square root of 4.9 is 2.21. Notice that we have here the beginnings of another series in which it is also true that a group of nonzero digits recurs and in which every move of the decimal point in the square root is associated with a move of *two* places in the square. In this series, however, the digits are 2, 2, and 1 instead of 7.

The purpose of this rather extended discussion is to give a rationale for the first step in extracting the square root. Beginning at the decimal point, we group the digits into periods of two digits apiece. The number 7304.27 is grouped as follows: 73 04.27. The number 913075.4882 is grouped as

________________________________.

– – – – – – – – – – – – – – –

91 30 75.48 82

If there is an odd number of digits to the *left* of the decimal point, the leftmost period will contain only one digit; for example, 843.27 is grouped as 8 43.27. If there is an odd number of digits to the *right* of the decimal point, a zero is appended to complete the rightmost period; for example, 51.063 is grouped as 51.06 30. The number 362.145 is grouped as ______________.

\- - - - - - - - - - - - - - -

3 62.14 50

Group the following numbers: 3.89247 ______________; .0973 ______________; 6145000 ______________.

\- - - - - - - - - - - - - - -

3.89 24 70; .09 73; 6 14 50 00

For our first example let us extract the square root of the number 5314.26, accurate to three significant figures: $(70)^2 = 4900$ and $(80)^2 = 6400$. Since 5314.26 falls between 4900 and 6400, its square root will lie somewhere between 70 and 80. If we group 5314.26 by the method described above, we will have

______________.

\- - - - - - - - - - - - - - -

53 14.26

Reading from left to right, consider the first period in this number, which consists of the digits 5 and 3. The largest perfect square that is less than or equal to the number 53 is 49, which has a square root of 7. Therefore we write the number 7 directly above the period 53 as follows:

$$\frac{7\quad .\quad}{53\ 14.26}$$

Notice that this arrangement is in accordance with the decision made above, namely, that the square root of 5314.26 lies somewhere between 70 and 80.

To develop a rationale for the next steps we shall do some algebra. Let us represent the difference between the real square root and our first estimate, which is 70, by the symbol **x**. Thus we have

$$(70 + x)^2 = 5314.26$$

Expanding the square on the left side, we have

$$\underline{\qquad\qquad\qquad\qquad\qquad} = 5314.26.$$

- - - - - - - - - - - - - - -

$\mathbf{4900 + 140x + x^2}$

Now we have the equation $4900 + 140x + x^2 = 5314.26$. Transposing 4900 to the right side, we have

____.

- - - - - - - - - - - - - - -

$\mathbf{140x + x^2 = 5314.26 - 4900 = 414.26}$

Now we have $140x + x^2 = 414.26$. Factoring x from the left member gives ______________________________.

– – – – – – – – – – – – – – –

$x(140 + x) = 414.26$

Division of both sides of the equation by $(140 + x)$ gives

____

– – – – – – – – – – – – – – –

$$\mathbf{x = \frac{414.26}{140 + x}}$$

Thus we see that the difference between 70 and the real square root of 5314.26 is approximately 414.26 divided by 140.

The conclusion that we have just reached by algebra can be represented in a manner similar to the process of "long division." Recall that we had reached the following stage in our procedure:

```
 7    .
---------
53 14.26
```

Now we write the square of 7 below 53 and subtract:

```
 7    .
---------
53 14.26
49
--
 4
```

Then we bring down the numbers from the next period, 14, to form a "partial remainder," which is 414:

```
 7   .
------
53 14.26
49
--
 4 14
```

We concluded above that we need to add to 70 a number that is approximately 414.26 divided by 140. We double our first digit 7 and write this "trial divisor" to the left of and one space below the partial remainder:

```
      7   .
     ------
     53 14.26
     49
     --
      4 14
(14
```

The doubling occurs because of the "cross-product term" involved in the squaring of (70 + x). Now we temporarily ignore the last digit of the partial remainder and consider how many times 14 will "go into" 41. It will go at least 2 times; therefore we write the number 2 in three places: (a) above the next period in 5314.26, (b) to the left of the parentheses in the trial divisor, and (c) directly after the trial divisor. This gives

```
       7  2.
      ------
      53 14.26
      49
      --
       4 14
2(142)
```

The "completed divisor," 142, corresponds to (140 + x), and the product 2(142) corresponds to x(140 + x) above. Two times 142 equals __________.

\- - - - - - - - - - - - - - -

284

This product, 284, is the difference between $(70)^2$ and $(72)^2$. However, it is less than 414.26, which is the difference between $(70)^2$ and 5314.26, the number whose square root we are seeking. To obtain the part of the square still unaccounted for we subtract and bring down as follows:

```
               7  2.
              ________
              53 14.26
              49
              __
               4 14
2(142) =       2 84
               ____
               1 30 26
```

We can again apply algebra to estimate a quantity, y, the difference between 72 and the true square root:

$$(72 + y)^2 = 5314.26$$

Expanding the square on the left side, we have

______________________ = 5314.26

- - - - - - - - - - - - - - -

$5184 + 144y + y^2$

Transposing 5184 to the right side, we obtain

______________________.

- - - - - - - - - - - - - - -

$144y + y^2 = 130.26$

Factoring y from the left member gives

________________________________.

\- - - - - - - - - - - - - -

$$y(144 + y) = 130.26$$

Dividing both sides by $(144 + y)$ gives

____________________.

\- - - - - - - - - - - - - -

$$y = \frac{130.26}{144 + y}$$

Thus the difference between 72 and the real square root will be approximately 130.26 divided by 144.

The stage we had reached in our procedure was

```
                 7  2 .
               ---------
               53 14.26
               49
               --
                4 14
2(142) =        2 84
                ----
                1 30 26
```

Doubling 72 and writing it as above, we obtain

```
                 7  2 .
               ---------
               53 14.26
               49
               --
                4 14
2(142) =        2 84
                ----
                1 30 26
 (144
```

Now we temporarily ignore the last digit of the partial remainder and consider how many times 144 goes into 1302. This estimate is __________.

– – – – – – – – – – – – – – –

9

Now we write the number 9 in three places as follows:

```
                7  2 . 9
              ----------
               53 14.26
               49
               --
                4 14
2(142)    =     2 84
                ----
                1 30 26
9(1449)   =     1 30 41
                -------
```

Notice that the effect of affixing the 9 to 144 and multiplying by 9 is to make our estimate a little too high. This situation occurs occasionally but need cause no concern. We simply lower our estimate by one digit, to 8, and repeat the same steps:

```
                7  2 . 8
              ----------
               53 14.26
               49
               --
                4 14
2(142)    =     2 84
                ----
                1 30 26
8(1448)   =     1 15 84
                -------
```

At the outset we said that we wanted the square root to be accurate to three significant figures. To ensure this we must carry out calculations to *one more figure* than desired in the answer to be reported and then *round off*. To accomplish this we affix a pair of zeros to the square and, after subtracting, carry them down as follows:

```
                7  2 . 8
              -------------
               53 14.26 00
               49
               --
                4 14
2(142)    =     2 84
                ----
                1 30 26
8(1448)   =     1 15 84
                -------
                  14 42 00
```

Doubling the existing quotient as before, we have

```
               7  2 . 8
              ___________
              53 14.26 00
              49
              --
               4 14
2(142)   =     2 84
               ----
               1 30 26
8(1448)  =     1 15 84
               -------
                 14 42 00
 (1456
```

The trial divisor 1456 goes into 14420 how many times?

__________.

\- - - - - - - - - - - - - -

9

The number 9 has been written appropriately below, but the reader should perform the indicated multiplication.

```
                 7  2 . 8  9
                ______________
                53 14.26 00
                49
                --
                 4 14
2(142)   =       2 84
                 ----
                 1 30 26
8(1448)  =       1 15 84
                 -------
                   14 42 00
9(14569) =         ________
```

- - - - - - - - - - - - - -

131121

This product is less than 144200, so we can now examine our present square root estimate, 72.89. The last digit is greater than 5, so we round upward. Thus, to three significant figures, the square root of 5314.26 is 72.9.

For a second example we shall extract the square root of the number 763.417 to three significant figures. As discussed above, we must first group the number into periods of two digits apiece, proceeding in both directions from the decimal point. The number 763.417 should be grouped as ______________________.

- - - - - - - - - - - - - -

7 63.41 70

Notice that the left-most period contains only one digit and that a zero has been appended to the third digit to the right of the decimal point to complete the last period. Now we set up the grouped number in the standard form:

$$\overline{7\ 63.41\ 70}$$

Looking at the first period, 7, we observe that the largest perfect square that is less than or equal to 7 is __________.

\- - - - - - - - - - - - - -

4

Write the square root of 4 directly above the 7.

$$\overline{7\ 63.41\ 70}$$

\- - - - - - - - - - - - - -

$$\begin{array}{l} \mathbf{2\ \ \ .} \\ \hline \mathbf{7\ 63.41\ 70} \end{array}$$

Write the square of 2 below 7, subtract, and bring down the next period.

```
 2   .
---------
 7 63.41 70
```

- - - - - - - - - - - - - -

```
 2   .
---------
 7 63.41 70
 4
 -
 3 63
```

Double 2 and write the result in the appropriate place.

```
 2   .
---------
 7 63.41 70
 4
 -
 3 63
```

- - - - - - - - - - - - - -

```
     2   .
    ---------
     7 63.41 70
     4
     -
     3 63
(4
```

Temporarily ignoring the last digit of the partial remainder 363, we consider how many times 4 goes into 36. Our first estimate of the next digit therefore is 9. Write 9 in the three appropriate places below.

```
      2   .
      ---------
      7 63.41 70
      4
      -
      3 63
(4
```

- - - - - - - - - - - - - - -

```
      2  9.
      ---------
      7 63.41 70
      4
      -
      3 63
9(49) =
```

Nine times 49 equals __________. This result is (less than, more than) the partial remainder 363 (underline one of the two phrases in parentheses).

- - - - - - - - - - - - - - -

441; more than

Since 441 is more than the partial remainder, we take as a second estimate the next lower digit, 8. Eight times 48 equals ________, which is (less than, more than) the partial remainder 363.

- - - - - - - - - - - - - - -

384; more than

Since 384 is also more than the partial remainder, we take as a third estimate the digit 7. Seven times 47 equals __________, which is (less than, more than) the partial remainder 363.

- - - - - - - - - - - - - - -

329; less than

Since 329, the product of 7 and 47, is less than 363, we can proceed to change our working arrangement accordingly. Write the appropriate number in the appropriate places below and perform the required multiplication.

```
    2    .
  ----------
    7 63.41 70
    4
    -
    3 63
(4
```

- - - - - - - - - - - - - - -

```
        2  7.
       ----------
        7 63.41 70
        4
        -
        3 63
7(47) = 3 29
        ----
```

Now subtract and bring down the next period.

```
            2  7 .
           ----------
           7 63.41 70
           4
           -
           3 63
7(47) =    3 29
           ----
```

```
            2  7 .
           ----------
           7 63.41 70
           4
           -
           3 63
7(47) =    3 29
           ----
             34 41
```

Double 27 and write the result in the appropriate place.

```
          2  7.
          ---------
          7 63.41 70
          4
          -
          3 63
7(47)  =  3 29
          -----
            34 41
```

- - - - - - - - - - - - - -

```
          2  7.
          ---------
          7 63.41 70
          4
          -
          3 63
7(47)  =  3 29
          -----
            34 41
 (54
```

Temporarily ignoring the last digit of the new partial remainder 3441, consider how many times 54 goes into 344. Our first estimate is therefore __________.

- - - - - - - - - - - - - -

6

Write 6 in the three appropriate places below.

```
              2  7.
             ___________
             7 63.41 70
             4
             _
             3 63
   7(47)  =  3 29
             ____
               34 41
    (54
```

- - - - - - - - - - - - - - -

```
          2  7.6
         ___________
         7 63.41 70
         4
         _
         3 63
7(47)  = 3 29
         ____
           34 41
6(546) =
```

Six times 546 equals ____________, which is (less than, more than) the partial remainder 3441.

- - - - - - - - - - - - - - -

3276; less than

Since the product is less than the partial remainder, write it below, subtract, and bring down the next period.

```
              2  7 . 6
             ___________
             7 63.41 70
             4
             _
             3 63
7(47)     =  3 29
             ____
               34 41
6(546) =
```

- - - - - - - - - - - - - - -

```
              2  7 . 6
             ___________
             7 63.41 70
             4
             _
             3 63
7(47)     =  3 29
             ____
               34 41
6(546) =       32 76
               _____
                1 65 70
```

Double 276 and write the result in the appropriate place.

```
               2  7 . 6
               ________
               7 63.41 70
               4
               _
               3 63
7(47)     =    3 29
               ____
                 34 41
6(546) =         32 76
                 _____
                  1 65 70
```

- - - - - - - - - - - - - - -

```
          2  7 . 6
          ________
          7 63.41 70
          4
          _
          3 63
7(47)  =  3 29
          ____
            34 41
6(546) =    32 76
            _____
             1 65 70
 (552
```

Temporarily ignoring the last digit of the new partial remainder 16570, consider how many times 552 goes into 1657. Our first estimate therefore is __________.

- - - - - - - - - - - - - - -

3

Write 3 in the three appropriate places below.

```
                2  7 . 6
               ___________
               7 63.41 70
               4
               _
               3 63
   7(47)    =  3 29
               ____
                 34 41
   6(546) =      32 76
                 _____
                  1 65 70
   (552
```

- - - - - - - - - - - - - - -

```
            2  7 . 6  3
           _____________
            7 63.41 70
            4
            _
            3 63
7(47)     = 3 29
            ____
              34 41
6(546)    =   32 76
              _____
               1 65 70
3(5523) =
```

Three times 5523 equals ____________, which is (less than, more than) the partial remainder 16570.

- - - - - - - - - - - - - - -

16569; less than

Since the product is less than the partial remainder, we can now examine the four figures in our present square root estimate, 27.63. Since the last digit is less than 5, it can be dropped. Therefore our final answer for the square root of 763.417 to three significant figures is ____________.

- - - - - - - - - - - - - -

27.6

For a third example let us extract the square root of the number 10.7228 to three significant figures. Grouping this number into periods, we have ________________________.

- - - - - - - - - - - - - -

10.72 28

Each of the three periods will give a significant figure in the square root. However, we wish accuracy to three significant figures, which requires that we obtain *four* figures and then round off. Therefore we append another period consisting of two zeros to our number 10.72 28, thus giving ________________________.

- - - - - - - - - - - - - -

10.72 28 00

Now determine the first digit in the square root, square, subtract, and bring down the next period, thus obtaining the first partial remainder.

```
  .
----------
10.72 28 00
```

- - - - - - - - - - - - - - -

```
 3 .
-----------
10.72 28 00
 9
--
 1 72
```

Carry the work further to the point of obtaining the second partial remainder.

```
 3 .
-----------
10.72 28 00
 9
--
 1 72
```

- - - - - - - - - - - - - - -

```
            3 . 2
           -----------
           10.72 28 00
            9
           --
            1 72
2(62) =     1 24
           -----
              48 28
```

Carry the work to the point of obtaining the third partial remainder.

```
            3 . 2
          ----------------
          10.72 28 00
           9
          --
           1 72
2(62) =    1 24
          ------
             48 28
```

- - - - - - - - - - - - - -

```
            3 . 2   7
          ----------------
          10.72 28 00
           9
          --
           1 72
2(62)  =   1 24
          ------
             48 28
7(647) =     45 29
            ------
              2 99 00
```

Obtain a new product and, if it is less than the partial remainder 29900, stop.

```
            3 . 2   7
           ___________
           10.72 28 00
            9
           __
            1 72
2(62)   =   1 24
           _____
              48 28
7(647)  =     45 29
              _____
               2 99 00
```

- - - - - - - - - - - - - - -

```
            3 . 2   7   4
           ______________
           10.72 28 00
            9
           __
            1 72
2(62)    =  1 24
           _____
              48 28
7(647)   =    45 29
              _____
               2 99 00
4(6544)  =     2 61 76
               _______
```

Thus the square root of 10.7228, rounded to three significant figures, is ____________.

- - - - - - - - - - - - - - -

3.27

For a fourth example extract the square root of 145,639.02 to three significant figures. Grouping this number into periods, we have ______________________.

- - - - - - - - - - - - - - -

14 56 39.02

Determine the first digit in the square root, square, subtract, and bring down the next period, thus obtaining the first partial remainder.

$$\overline{14\ 56\ 39.02}$$

- - - - - - - - - - - - - - -

3 .

14 56 39.02

9

5 56

Carry the work to the point of obtaining the second partial remainder.

```
  3      .
 ----------
 14 56 39.02
  9
 --
  5 56
```

- - - - - - - - - - - - - - -

```
           3  8     .
          -----------
          14 56 39.02
           9
          --
           5 56
8(68) =    5 44
          -----
             12 39
```

Carry the work to the point of obtaining the third partial remainder.

```
            3  8    .
          __________
          14 56 39.02
           9
          __
           5 56
8(68) =    5 44
           ____
             12 39
```

- - - - - - - - - - - - - - -

```
              3  8  1.
            __________
**          14 56 39.02
             9
            __
             5 56
8(68)   =    5 44
             ____
               12 39
1(761)  =       7 61
              ______
                4 78 02
```

Obtain a new product and, if it is less than the partial remainder 47802, stop.

```
                 3  8  1.
               ___________
               14 56 39.02
                9
               __
                5 56
8(68)    =      5 44
               _____
                 12 39
1(761)   =        7 61
                 _____
                  4 78 02
```

- - - - - - - - - - - - - - -

```
                 3  8  1.6
               ___________
               14 56 39.02
                9
               __
                5 56
8(68)    =      5 44
               _____
                 12 39
1(761)   =        7 61
                 _____
                  4 78 02
6(7626)  =        4 57 56
                  _______
```

Thus the square root of 145,639.02, rounded to three significant figures, is __________.

- - - - - - - - - - - - - - -

382

REVIEW TEST

Extract the square root of these numbers, accurate to two significant figures:

1. 637.85
2. 70.24
3. .0294
4. 4581

Extract the square root of these numbers, accurate to three significant figures:

5. 181,697.35
6. 74.45

Chapter IV. Use of the Summation Operator and Summation Laws

OBJECTIVES

Upon completion of this chapter, the reader should be able to use the following:

1. The summation operator to express the sum of a series of scores.

2. The summation operator to express the sum of a series of squares of scores.

3. The three summation laws to change algebraic expressions into a form more convenient for statistical computations.

THE SUMMATION OPERATOR

In statistics we are concerned with a series of values assumed by a variable. Let us call these values "scores." It is often useful to combine the scores or to combine simple functions of them. For this purpose we use as a mathematical operator the capital Greek letter "sigma," which is denoted as "Σ." This operator can be translated as "the sum of."

Suppose that we have three scores, 7, 4, and 6. The sum of these scores is __________.

- - - - - - - - - - - - - - - -

17

Assume that we choose to represent the underlying variable as X. To indicate symbolically the sum of the three scores we can write ΣX. In this example the variable X takes on three different values. We use the letter N to indicate the number of scores to be summed. In this example $N =$ __________.

- - - - - - - - - - - - - - - -

3

To express the example in a more complete form we can write $\sum^{3} X = 7 + 4 + 6 = 17$. In practice we do not need to write out the intermediate expression but rather can write $\sum^{N} X = 17$. However, the foregoing expression makes explicit the fact that there are in this example three scores to be added and that the three scores are __________, __________, and __________.

- - - - - - - - - - - - - - -

7; 4; 6

In many applications of statistics N represents the number of persons being measured on a given trait, and each score represents the measurement for a particular person on this trait. In the more elementary applications the series of persons represents the only dimension being dealt with. In more advanced applications, however, a second dimension may be involved; for example, days of the week. Suppose that each of the three persons is measured on Monday, Tuesday, Wednesday, and Thursday. Then for each of our three persons we would have not one but __________ scores.

- - - - - - - - - - - - - - -

4

The following table shows four scores for each of three persons.

Person	Monday	Tuesday	Wednesday	Thursday
1	7	3	4	9
2	4	8	6	1
3	6	1	3	4

We might choose to represent the number of days by the letter D. In this example $D =$ ________.

- - - - - - - - - - - - - - -

4

Suppose that we wish to represent the sum of the scores on the four days for a single person, namely, Person 1. The four scores for Person 1 are ________, ________, ________, and ________ and their sum is ________.

- - - - - - - - - - - - - - -

7; 3; 4; 9; 23

In an analogous manner we can represent this sum as $\overset{4}{\Sigma} X = 7 + 3 + 4 + 9 = 23$ or more simply as $\overset{D}{\Sigma} X = 23$. In this example we had to make a distinction between a sum over *persons* and a sum over *days*. To do so we wrote a letter over the summation sign. The letter we used for the number of persons was ________, and the letter we used for the number of days was ________.

- - - - - - - - - - - - - - -

N; D

When we deal with only one dimension, there is no ambiguity, and we can write the summation sign without any letter above it.

The summation sign can be used to indicate not only the sum of a series of scores but also the sum of a particular function of the scores. The most frequent example is to indicate the sum of the *squares* of the scores. Consider the scores in the foregoing example for the three persons on Monday, 7, 4, and 6. In expressing variability of scores it is useful to square the scores and then sum them. The square of 7 is 49, the square of 4 is ______, and the square of 6 is ______.

- - - - - - - - - - - - - - -

16; 36

The sum of 49, 16, and 36 is __________.

- - - - - - - - - - - - - - -

101

Symbolically, we can express these operations as $\sum^{3} X = (7)^2 + (4)^2 + (6)^2 = 49 + 16 + 36 = 101$ or simply as $\sum^{N} X^2 = 101$. The value of $\sum^{N} X^2$ for Monday is 101. To find the value of $\sum^{N} X^2$ for Tuesday we must square each of the scores for the three persons on Tuesday, which are 3, 8, and 1. The squares are, respectively, __________, __________, and __________.

- - - - - - - - - - - - - - -

9; 64; 1

Therefore the sum of the *squares* of the scores on Tuesday is __________.

- - - - - - - - - - - - - - -

74

The value of $\sum^{N} X^2$ for Tuesday is ________.

- - - - - - - - - - - - - - -

74

Calculate the value of $\sum^{N} X^2$ for Wednesday. The answer is ________.

- - - - - - - - - - - - - - -

61

The value of $\sum^{N} X^2$ for Thursday is ________.

- - - - - - - - - - - - - - -

98

Rather than summing the squares of scores over *persons*, we can sum them over *days*. For Person 1 we have

$$\sum^{4} X^2 = (7)^2 + (3)^2 + (4)^2 + (9)^2$$

$$= 49 + 9 + 16 + 81 = 155$$

or more simply

$$\sum^{D} X^2 = 155$$

For Person 2 we have

$$\sum^{4} X^2 = (4)^2 + (8)^2 + (6)^2 + (1)^2$$

$$= \_\_\_\_\_ + \_\_\_\_\_ + \_\_\_\_\_ + \_\_\_\_\_ = \_\_\_\_\_$$

or more simply

$$\sum^{D} X^2 = \_\_\_\_\_\_$$

- - - - - - - - - - - - - -

16; 64; 36; 1; 117; 117

Calculate the value of $\sum^{D} X^2$ for Person 3. The answer is

__________.

- - - - - - - - - - - - - -

62

We have computed the value of $\overset{N}{\Sigma}X$ for Monday to be $7 + 4 + 6$ or 17. In certain statistical procedures we wish to indicate the *square of a sum.* In this case the square of the sum would be $(17)^2 = 289$. Symbolically, it is represented as $(\overset{N}{\Sigma}X)^2 = 289$.

Notice very carefully that the "square of the sum" is in general *not* equal to the "sum of the squares." For the present series of scores the sum of the squares is

$$\overset{N}{\Sigma}X^2 = (7)^2 + (4)^2 + (6)^2 = 49 + 16 + 36 = 101$$

101 is *not* equal to 289. The difference lies in the "order of operations." In the case of $(\overset{N}{\Sigma}X)^2$ first we sum and then we square. In the case of $\overset{N}{\Sigma}X^2$ first we square and then we sum.

A more complete way of indicating the sum of the squares would be $\overset{N}{\Sigma}(X)^2$. The parentheses clearly distinguish the order of operations as different from that in $(\overset{N}{\Sigma}X)^2$. However, conventional usage has established $\overset{N}{\Sigma}X^2$ as meaning $\overset{N}{\Sigma}(X)^2$, so we shall continue to use the simpler representation.

For Tuesday we have

$$(\overset{N}{\Sigma}X)^2 = (3 + 8 + 1)^2 = (12)^2 = \_\_\_\_\_\_\_\_\_\_$$

144

On the other hand, for Tuesday

$$\sum^{N} X^2 = (3)^2 + (8)^2 + (1)^2 = 9 + 64 + 1 = \_\_\_\_\_\_\_\_\_\_ .$$

- - - - - - - - - - - - - - -

74

Notice that for Tuesday also $(\sum^{N} X)^2 = 144$ is *not* equivalent to $\sum^{N} X^2 = 74$. Calculate $(\sum^{N} X)^2$ for Wednesday. The answer is

__________ .

- - - - - - - - - - - - - - -

169

The value of $(\sum^{N} X)^2$ for Thursday is __________ .

- - - - - - - - - - -

196

Now let us again deal with summations over *days*. For Person 1 compute all three quantities, $\overset{D}{\Sigma}X$, $(\overset{D}{\Sigma}X)^2$, and $\overset{D}{\Sigma}X^2$. The answers are, respectively, ________, ________, and ________.

- - - - - - - - - - - - - - -

23; 529; 155

For Person 2, $\overset{D}{\Sigma}X =$ __________, $(\overset{D}{\Sigma}X)^2 =$ __________, and $\overset{D}{\Sigma}X^2 =$ __________.

- - - - - - - - - - - - - - -

19; 361; 117

FIRST SUMMATION LAW

In dealing with the more complex statistical procedures we usually find that the formula that most directly shows the theoretical rationale is *not* convenient for computational purposes. Therefore so-called "raw-score formulas" are often derived. In the derivation of these formulas three summation laws are of great utility. These three laws are also used to develop further theoretical implications.

In the exposition below we shall use small computational examples to make the summation laws seem intuitively plausible. It may seem to the reader that, in these small examples, the laws don't involve much of a saving. However, he should try to imagine what their effect would be if the number of scores were quite large.

To develop the first summation law we shall use the same "data matrix" as was used above. For convenience of reference, it is repeated below.

Person	Monday	Tuesday	Wednesday	Thursday
1	7	3	4	9
2	4	8	6	1
3	6	1	3	4

Suppose that we wish to obtain the "grand total" of scores for all persons on all days. We could proceed either (a) by first summing the *rows* to obtain person totals and then adding the three person totals or (b) by first summing the *columns* to obtain day totals and then adding the four day totals. Let us first proceed by obtaining person totals. The total for Person 1 is _____.

\- - - - - - - - - - - - - - -

23

The total for Person 2 is __________ and for Person 3 is __________.

- - - - - - - - - - - - - - -

19; 14

Now we have three person totals, 23, 19, and 14. Their grand total is __________.

- - - - - - - - - - - - - - -

56

Now let us obtain the day totals. The day total for Monday is __________.

- - - - - - - - - - - - - - -

17

The day total for Tuesday is __________, for Wednesday __________, and for Thursday __________.

\- - - - - - - - - - - - - - - -

12; 13; 14

Thus we have the four day totals, 17, 12, 13, 14. Adding these, we obtain __________.

\- - - - - - - - - - - - - - - -

56

Notice that the same grand total of 56 is obtained by adding either the row totals or the column totals. Now let us use the following letters to represent the raw scores as variables.

W represents a score on Monday.
X represents a score on Tuesday.
Y represents a score on Wednesday.
Z represents a score on Thursday.

For Person 1 $W = 7$, $X = 3$, $Y = 4$, and $Z = 9$. Therefore for Person 1

$$(W + X + Y + Z) = (7 + 3 + 4 + 9) = \_\_\_\_\_\_\_\_.$$

- - - - - - - - - - - - - - -

23

For Person 2 $(W + X + Y + Z) = (4 + 8 + 6 + 1) = \_\_\_\_\_\_\_\_.$

- - - - - - - - - - -

19

For Person 3 $(W + X + Y + Z) = ($__________ + __________ + ________ + ________$) =$ ________.

- - - - - - - - - - - - - - - -

6; 1; 3; 4; 14

Now suppose that we wish to indicate the grand total as a sum of these row totals. We can write $\overset{N}{\Sigma}(W + X + Y + Z) = 23 + 19 + 14 = 56$.

Notice that the summation here is over the *rows;* hence the letter N, which equals the number of rows, 3.

It was shown above, however, that the same grand total can be obtained by adding the *column* totals, that is, $17 + 12 + 13 + 14 = 56$. How can we indicate symbolically the first column total? We can write $\overset{N}{\Sigma}W$. This means that we should take the three values of W (representing the score on Monday), which are 7, 4, and 6, and add them; $\overset{N}{\Sigma}W =$ __________.

- - - - - - - - - - - - - - - -

17

To get the day total for Tuesday we take $\sum^{N} X =$ ________.

- - - - - - - - - - - - - -

12

$\sum^{N} Y =$ ________ and $\sum^{N} Z =$ ________

- - - - - - - - - - - - - -

13; 14

In summary we have

$$\sum^{N}(W + X + Y + Z) = \sum^{N}W + \sum^{N}X + \sum^{N}Y + \sum^{N}Z$$

$$\sum^{N}(W + X + Y + Z) = 17 + 12 + 13 + 14 = 56$$

Stating this law in more abstract terms, we can say, "The summation sign can be distributed to each of a set of variables that are to be added." Actually, this law can be stated more generally. It is also true that

$$\sum^{N}(W + X - Y) = \sum^{N}W + \sum^{N}X - \sum^{N}Y$$

To see that this equation is plausible consider the following example.

	W	X	Y	$(W + X - Y)$
	2	7	3	6
	5	9	0	14
	0	4	1	3
$\sum^{N}$	7	20	4	23

In this example we use the word "combine" to indicate either addition or subtraction. In the first row we combine as follows: $2 + 7 - 3 = 6$. In the second row, $5 + 9 - 0 =$ __________, and in the third row, $0 + 4 - 1 =$ __________.

14; 3

The grand total of 23 can be obtained by summing the row totals:

$$6 + 14 + 3 = 23$$

The same grand total can be obtained by *combining* the column totals:

$$\sum^{N}(W + X - Y) = \sum^{N}W + \sum^{N}X - \sum^{N}Y$$

$$\sum^{N}(W + X - Y) = 7 + 20 - 4 = 23$$

Thus a more general statement of the first summation law is, "The summation sign can be distributed to each of a set of variables that are to be combined." In many statistical procedures it is more accurate to get the column sums first and then to combine them in some specified manner.

SECOND SUMMATION LAW

Suppose that we wish to multiply each of a series of numbers by the same number (i.e., a "constant") and then add the products. Here is an example:

k	X	kX
6	5	30
6	−3	−18
6	0	0
6	8	48

The sum of the products is (30) + (−18) + (0) + (48) = ______.

\- - - - - - - - - - - - - - -

60

We can obtain this same sum by adding the four values of the variable X and then multiplying *this sum* by the constant k. Combining the four values of X, we have

$$\sum^{N} X = (5) + (-3) + (0) + (8) = \_\_\_\_\_\_\_$$

\- - - - - - - - - - - - - - -

10

Multiplying the sum 10 by the constant 6, we have ________.

- - - - - - - - - - - - - - -

60

Notice that this is the same numerical result as obtained by the earlier procedure. In symbolic form we have

$$\sum^{N} kX = k\sum^{N} X$$

$$(30) + (-18) + (0) + (48) = 6(10)$$

This process is usually called taking a constant factor "outside" the summation sign. The second summation law can be stated as follows: "A constant factor can be taken outside the summation sign"; for example,

$\Sigma 13Y = 13\Sigma Y$; $\Sigma 5W =$ ____________

- - - - - - - - - - - - -

$5\Sigma W$

In the foregoing examples the constant was expressed as a number. In some cases, however, it is useful to express a constant as a letter. The convention has been to use letters from the *front* part of the alphabet to represent *constants*, whereas letters from the *back* part of the alphabet are used to represent *variables;* for example,

$$\Sigma bZ = b\Sigma Z$$

Here b is used as a constant and Z as a variable.

$$\Sigma gT = \underline{\qquad\qquad}$$

- - - - - - - - - - - - - - -

$g\Sigma T$

$\Sigma 3cZ = 3c\Sigma Z$. Here 3 and c are each constants and both are taken outside the summation sign.

$\Sigma 7XY = 7\Sigma XY$. Here X and Y are each variables and both are left "inside" the summation sign.

$\Sigma 19eUV = \underline{\qquad\qquad\qquad}$

- - - - - - - - - - - - - - -

$19e\Sigma UV$

$\Sigma 67Q^2 = 67\Sigma Q^2$. Notice that the square of a variable is handled in the same way as the corresponding variable. Similarly, in the case of the square of a constant

$$\Sigma a^2 S = a^2 \Sigma S; \quad \Sigma f^2 y^2 = f^2 \Sigma y^2; \quad \Sigma c^2 x = \_\_\_\_\_\_\_\_$$

- - - - - - - - - - - - - - -

$c^2 \Sigma x$

$\Sigma 28ay^2 z = \_\_\_\_\_\_\_\_\_\_\_\_$

- - - - - - - - - - - - - - -

$28a\Sigma y^2 z$

Now let us see how the first and second summation laws can be used together. Consider the expression $\Sigma(x^2 - 2xy + y^2)$. Many readers will recognize the expression in parentheses as the square of the binomial $(x - y)$. Initially we can use the first summation law to distribute the summation sign to the three terms. We have

$$\Sigma(x^2 - 2xy + y^2) = \Sigma x^2 - \Sigma 2xy + \Sigma y^2$$

Next consider separately the second of the three terms on the right side of the equation, $\Sigma 2xy$. It consists of the sum of a constant times two variables. Hence we can use the second summation law on this term:

$\Sigma 2xy =$ ____________________

\- - - - - - - - - - - - - - - -

$2\Sigma xy$

Inserting $2\Sigma xy$ into the equation above in place of $\Sigma 2xy$, we have

$$\Sigma(x^2 - 2xy + y^2) = \Sigma x^2 - 2\Sigma xy + \Sigma y^2$$

Apply the first and second summation laws to this equation:

$\Sigma(T^2 + 4S) =$ __________________________

\- - - - - - - - - - - - - - - -

$\Sigma T^2 + 4\Sigma S$

Apply the first and second summation laws to this equation:

$\Sigma(dY^2 - 4X + eWZ - V) =$ ______________________.

- - - - - - - - - - - - - - -

$\boldsymbol{d\Sigma Y^2 - 4\Sigma X + e\Sigma WZ - \Sigma V}$

THIRD SUMMATION LAW

So far we have considered the sum of a variable and the sum of a constant times a variable. How about the sum of a constant by itself? To learn how to deal with this third situation consider the following example:

Person	X
1	3
2	3
3	3
4	3
5	3

In this example each of five persons obtained exactly the same score, namely, 3, on some variable X. We wish to obtain the sum of their scores. We could add the scores serially, that is, $3 + 3 = 6$, $6 + 3 = 9$, $9 + 3 = 12$, and $12 + 3 = 15$. However, the reader can probably think of a simpler procedure. The procedure is to multiply the constant score by the number of times it occurs (in this example by the number of persons). There are 5 persons and the constant score is 3; therefore 5 times 3 equals

__________.

- - - - - - - - - - - - - - -

15

Verbally, the third summation law can be stated: "The sum of a constant equals the constant times the number of units over which it is summed." Symbolically, it can be stated

$$\sum^{N} k = kN$$

for example, $\sum^{N} 8 = 8N$ and $\sum^{d} 14 = 14d$. $\sum^{h} 31 =$ ____________.

- - - - - - - - - - - - - -

31*h*

The main use to which the third summation law is put involves situations in which the constant is expressed by a letter rather than a number; for example,

$$\sum^{N} b = bN \text{ and } \sum^{f}(-a^2) = -a^2f \quad \sum^{N} gh = \_\_\_\_\_\_\_\_\_\_\_\_$$

- - - - - - - - - - - - - -

ghN

$$\sum^{m}(-6ef^2) = \_\_\_\_\_\_\_\_\_\_\_\_\_\_\_\_\_\_\_\_$$

- - - - - - - - - - - - -

$-6ef^2m$

The constants involving letters may be arranged in any order. In the preceding example $-6mef^2$ would have been an equally correct answer and so would $-6emf^2$. However, the convention is to write constants involving numbers before constants involving letters, and $-m6ef^2$, although technically correct, is not considered an acceptable form.

Now let us employ all three summation laws to simplify the expression $\sum^N(7UV - 3j)$. Initially we can distribute the summation sign by using the first summation law:

$$\sum^N(7UV - 3j) = \underline{\hspace{6cm}}$$

- - - - - - - - - - - - - - -

$\sum^N 7UV - \sum^N 3j$

We shall handle the two terms separately. The first term involves the product of a constant and two variables. Applying the second summation law, we have

$$\sum^N 7UV = \underline{\hspace{4cm}}$$

- - - - - - - - - - - - - - -

$7\sum^N UV$

The second term involves the product of two constants. Applying the third summation law, we have

$$\sum^{N} 3j = \_\_\_\_\_\_\_\_\_\_\_\_$$

- - - - - - - - - - - - - - -

$3Nj$ [or] **$3jN$**

Inserting these two expressions into the equation above, we have

$$\sum^{N}(7UV - 3j) = \_\_\_\_\_\_\_\_\_\_\_\_$$

- - - - - - - - - - - - - - -

$7\sum^{N} UV - 3Nj$

Notice that the summation sign appears in the first term but *not* in the second term.

In the preceding example the number of units over which the summation is intended was indicated explicitly. In many applications in introductory statistics, however, the number of units is clear from the context, and the summation sign is written by itself. In the example if the context had clearly indicated that the summation was over N units the steps could have been written as

$$\begin{aligned}\Sigma(7UV - 3j) &= \Sigma 7UV - \Sigma 3j \\ &= 7\Sigma UV - 3Nj\end{aligned}$$

Apply the summation laws to the expression below, in which the summation is over m units.

$$\Sigma(19c^2 + 4x - yz) = \_\_\_\_\_\_\_\_\_\_\_\_\_\_\_\_.$$

- - - - - - - - - - - - - - -

$19mc^2 + 4\Sigma x - \Sigma yz$

[The first term may be written **$19c^2m$**.]

Let us review this problem before going on. Consider the terms within the parentheses separately. The first term $19c^2$ consists entirely of constant quantities. Therefore the third summation law is applicable, and the first term of the answer $19mc^2$ does *not* contain a summation sign. Rather, the number of units over which summed, m, has been introduced as a factor.

Now consider the second term in parentheses, $4x$. This term constitutes a constant 4 and a variable x. Therefore the second summation law is applicable, and the corresponding term in the answer is the constant times the sum of the variable; that is, $4\Sigma x$.

The third term consists of the product of two variables. Therefore the first summation law is applicable. The third term of the answer is the sum of the product of the variables.

Apply the summation laws to the expression below, in which the summation is over k units.

$$\Sigma(abst - 8W^2y + 21F^2) = \underline{\hspace{6cm}}$$

- - - - - - - - - - - - - - -

$\mathbf{ab\Sigma st - 8\Sigma W^2y + 21kF^2}$
[The last term may be written $\mathbf{21F^2k}$.]

Apply the summation laws to the expression below, in which the summation is over c units.

$$\Sigma(12Agt + 3d - rwz + h^2XY) =$$

____

\- - - - - - - - - - - - - - -

$\mathbf{12Ag\Sigma t + 3cd - \Sigma rwz + h^2\Sigma XY}$

[The second term may be written **$3dc$**.]

Finally, let us collect in compact form verbal expressions of the three summation laws:

First Summation Law. The summation sign can be distributed to each of a set of variables that are to be combined.

Second Summation Law. A constant factor can be taken outside the summation sign.

Third Summation Law. The sum of a constant equals the constant times the number of units over which it is summed.

REVIEW TEST

For the first 10 problems use the following data matrix:

Person	Month		
	April	May	June
1	5	4	7
2	8	1	6
3	11	9	6
4	4	13	2
5	3	2	10

Let the variable being measured be represented by X, let N represent the number of persons, and let M represent the number of months.

1. $N =$

2. $M =$

3. For Person 1 find $\sum^{M} X$, $(\sum^{M} X)^2$, and $\sum^{M} X^2$

4. For Person 2 find $\sum^{M} X$, $(\sum^{M} X)^2$, and $\sum^{M} X^2$

5. For Person 3 find $\sum^{M} X$, $(\sum^{M} X)^2$, and $\sum^{M} X^2$

6. For Person 4 find $\sum^{M} X$, $(\sum^{M} X)^2$, and $\sum^{M} X^2$

7. For Person 5 find $\sum^{M} X$, $(\sum^{M} X)^2$, and $\sum^{M} X^2$

8. For April, find $\sum^{N} X$, $(\sum^{N} X)^2$, and $\sum^{N} X^2$

9. For May, find $\sum^{N} X$, $(\sum^{N} X)^2$, and $\sum^{N} X^2$

10. For June, find $\sum^{N} X$, $(\sum^{N} X)^2$, and $\sum^{N} X^2$

In each of the remaining problems use the three summation laws to change the algebraic expression into a form more convenient for statistical computations.

11. $\Sigma(S + T) =$

12. $\Sigma(u + v - w) =$

13. $\Sigma 6Y =$

14. $\Sigma cX =$

15. $\Sigma 23k^2 z =$

16. $\Sigma 12VW =$

17. $\Sigma 8Ay^2 z =$

18. $\Sigma(-15D^2 gs^2 T) =$

19. $\overset{N}{\Sigma} 9 =$

20. $\overset{N}{\Sigma} d =$

21. $\overset{b}{\Sigma} e =$

22. $\overset{h}{\Sigma} 17 =$

In Problems 23–30 assume that the summation is over N units.

23. $\Sigma(x - 14Z) =$

24. $\Sigma(5W + 8Y^2) =$

25. $\Sigma(aU - 4V + 7jW) =$

26. $\Sigma(e^2X + 9C^2y - 26aFyz) =$

27. $\Sigma(s - 8) =$

28. $\Sigma(15t^2 + D) =$

29. $\Sigma(3b - 9Wx + gLy) =$

30. $\Sigma(38As + cK^2T - 25b^2E) =$

ANSWERS FOR FIRST REVIEW TEST ON ALGEBRA

1. $\frac{7}{12}$

2. $\frac{13}{18}$

3. $\frac{1}{12}$

4. $\frac{11}{36}$

5. $\frac{3y + 7x}{xy}$

6. $\frac{n - 2k}{4kn}$

7. $\frac{8t - 9Ct + 6C}{12Ct}$

8. $\frac{8}{15}$

9. $\frac{70}{117}$

10. $\frac{6f}{35Bs}$

11. $\frac{28}{45}$

12. $\frac{14}{15}$

13. $\frac{63M}{2}$

14. 2

15. 6.08

16. 13.743

17. .00042

18. 70

19. 7.5

20. 5.8

21. 7

22. –28

23. 13.2

24. –28

25. 9.55

26. –19.344

27. 126.9

28. –6.4

29. 1.7

30. $-13bc/9d$

ANSWERS FOR SECOND REVIEW TEST ON ALGEBRA

1. 40
2. $11y$
3. $3L - 18 + 3Q$
4. 60
5. $43p - 21px$
6. $96z - 39y - 72$
7. 48
8. 0
9. $\sqrt{55w}$
10. $\sqrt{7}$
11. $\sqrt{15Ab}$
12. $\sqrt{54x^2 - 45x^3}$
13. $\sqrt{28c + 15}$
14. $\sqrt{\frac{S}{N} - \frac{T^2}{N^2}}$ [or] $\sqrt{\frac{S}{N} - \left(\frac{T}{N}\right)^2}$
15. .8
16. .15
17. 5
18. 3/7
19. −7
20. 5/24
21. $6c - 3$ [or] $3(2c - 1)$
22. $9 - 14d/3b$
23. $4 - 6a + 9a^2/4$
24. 5
25. $\frac{S(Y - N)}{T} + M$

ANSWERS FOR REVIEW TEST ON PLOTTING POINTS AND LINEAR EQUATIONS ON GRAPHS

1.

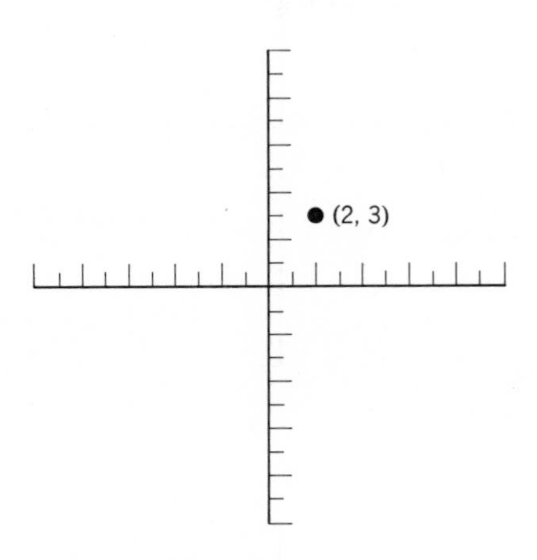

2.

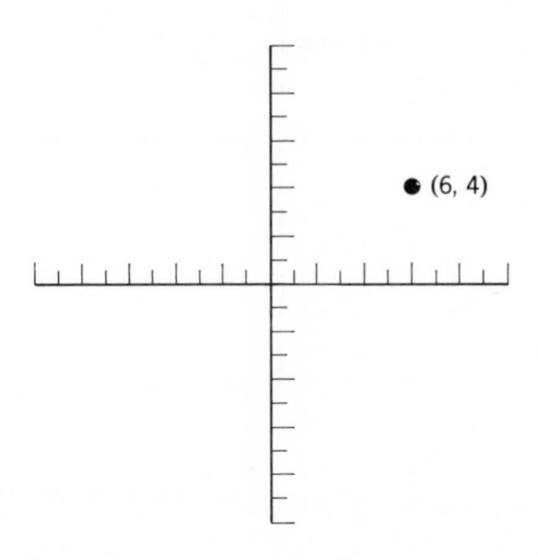

3.

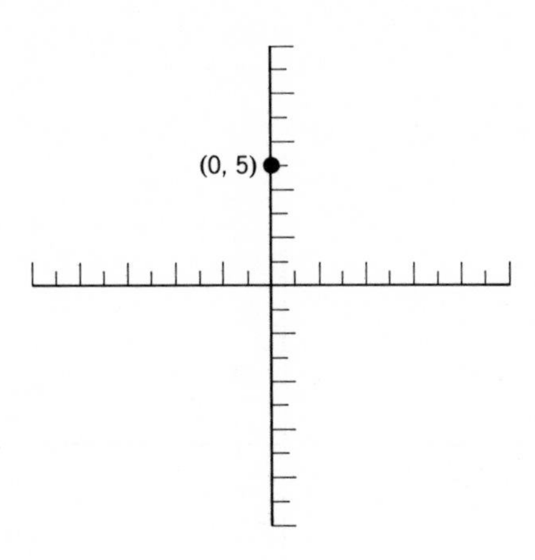

4.

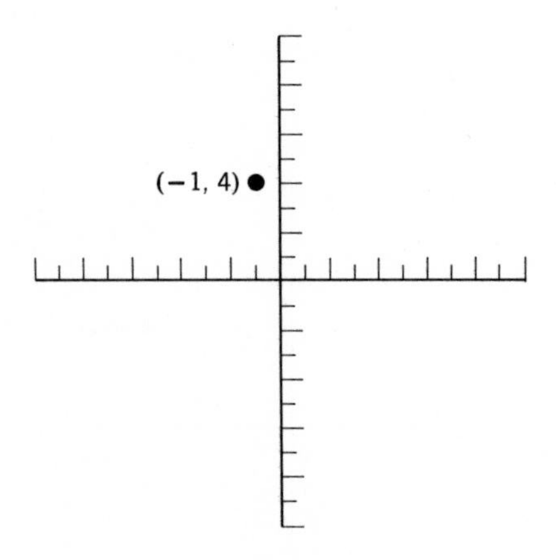

5.

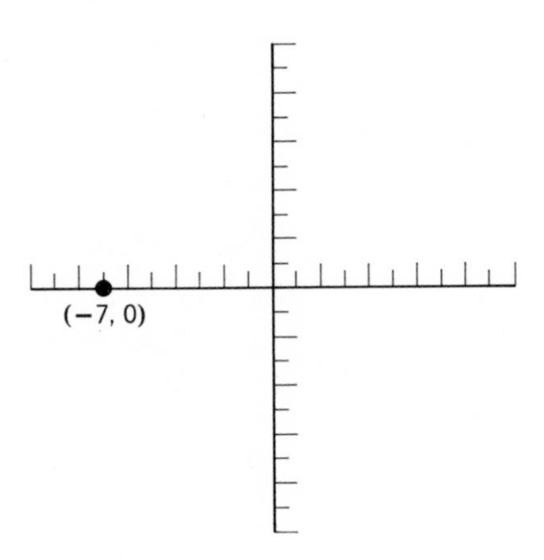

6.

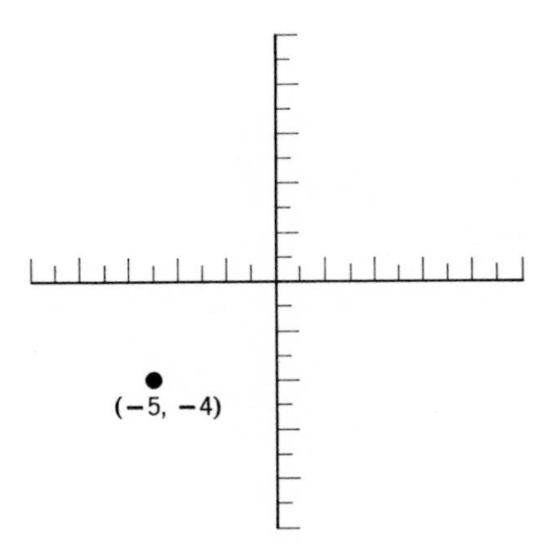

7.

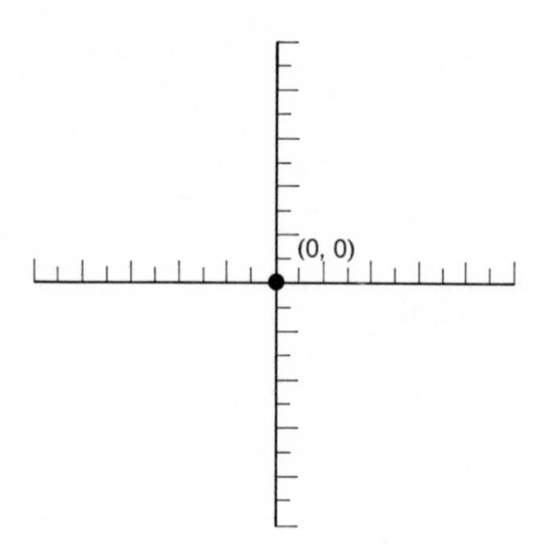

8.

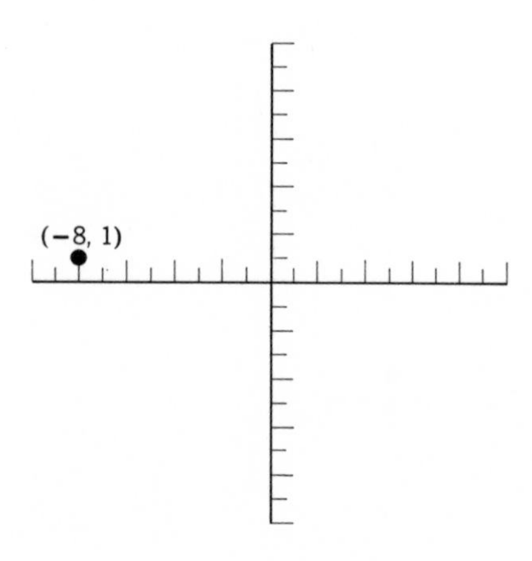

9.

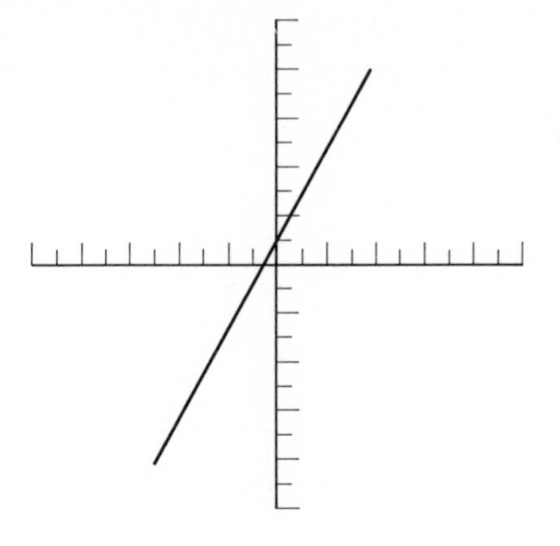

y intercept, 1 slope, 2

10.

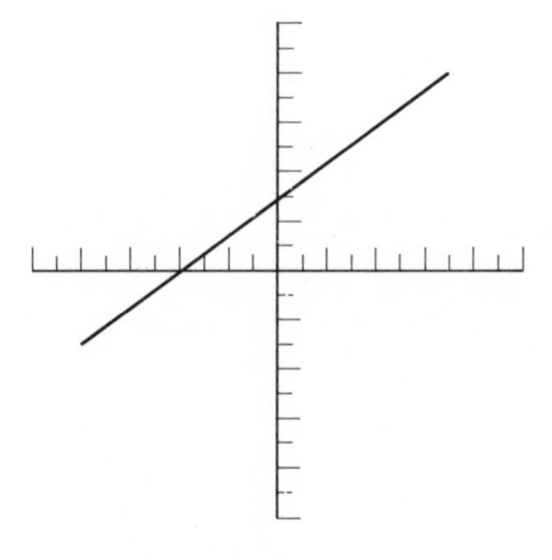

y intercept, 3 slope, 3/4

11.

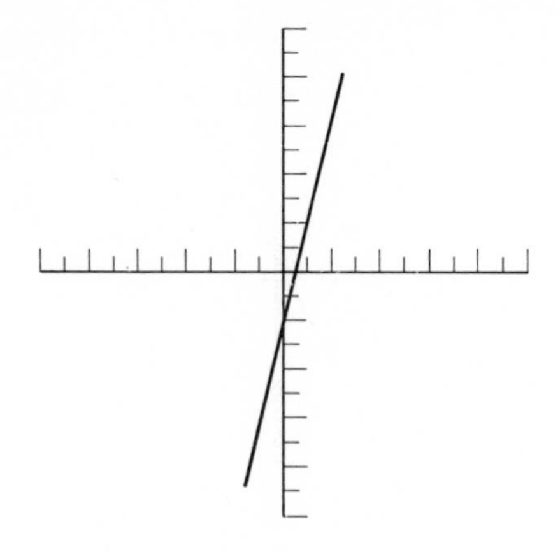

y intercept, -2 slope, 4

12.

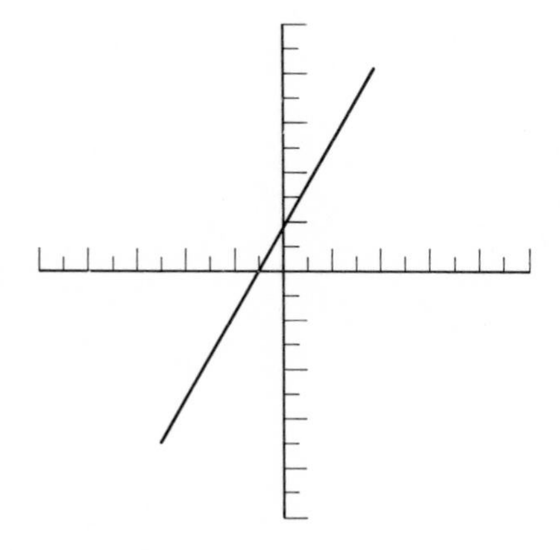

y intercept, 2 slope, 5/3

13.

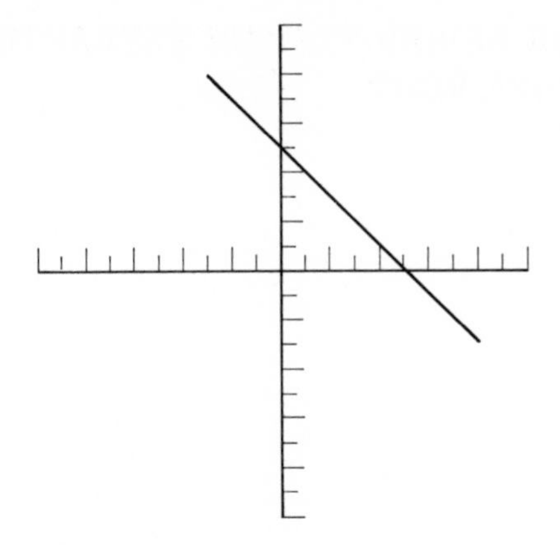

y intercept, 5 slope, −1

14.

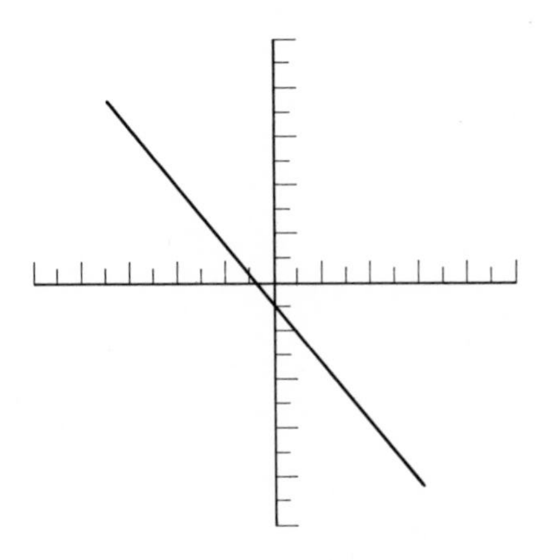

y intercept, −7/2 slope, −5/2

ANSWERS FOR REVIEW TEST ON EXTRACTION OF THE SQUARE ROOT

1. 25

2. 8.4

3. .17

4. 68

5. 426

6. 8.63

ANSWERS TO THE REVIEW PROBLEMS ON USE OF THE SUMMATION OPERATOR AND SUMMATION LAWS

1. 5

2. 3

3. 16; 256; 90

4. 15; 225; 101

5. 26; 676; 238

6. 19; 361; 189

7. 15; 225; 113

8. 31; 961; 235

9. 29; 841; 271

10. 31; 961; 225

11. $\Sigma S + \Sigma T$

12. $\Sigma u + \Sigma v - \Sigma w$

13. $6\Sigma Y$

14. $c\Sigma X$

15. $23k^2\Sigma z$

16. $12\Sigma VW$

17. $8A\Sigma y^2 z$

18. $-15D^2 g\Sigma s^2 T$

19. $9N$

20. dN [or] Nd

21. eb [or] be

22. $17h$

23. $\Sigma x - 14\Sigma Z$

24. $5\Sigma W + 8\Sigma Y^2$

25. $a\Sigma U - 4\Sigma V + 7j\Sigma W$

26. $e^2\Sigma X + 9C^2\Sigma y - 26aF\Sigma yz$

27. $\Sigma s - 8N$

28. $15\Sigma t^2 + DN$ [The last term may be ND.]

29. $3bN - 9\Sigma Wx + gL\Sigma y$ [The first term may be $3Nb$.]

30. $38A\Sigma s + cK^2\Sigma T - 25b^2EN$ [The last term may be $-25b^2NE$ or $-25Nb^2E$.]

Index

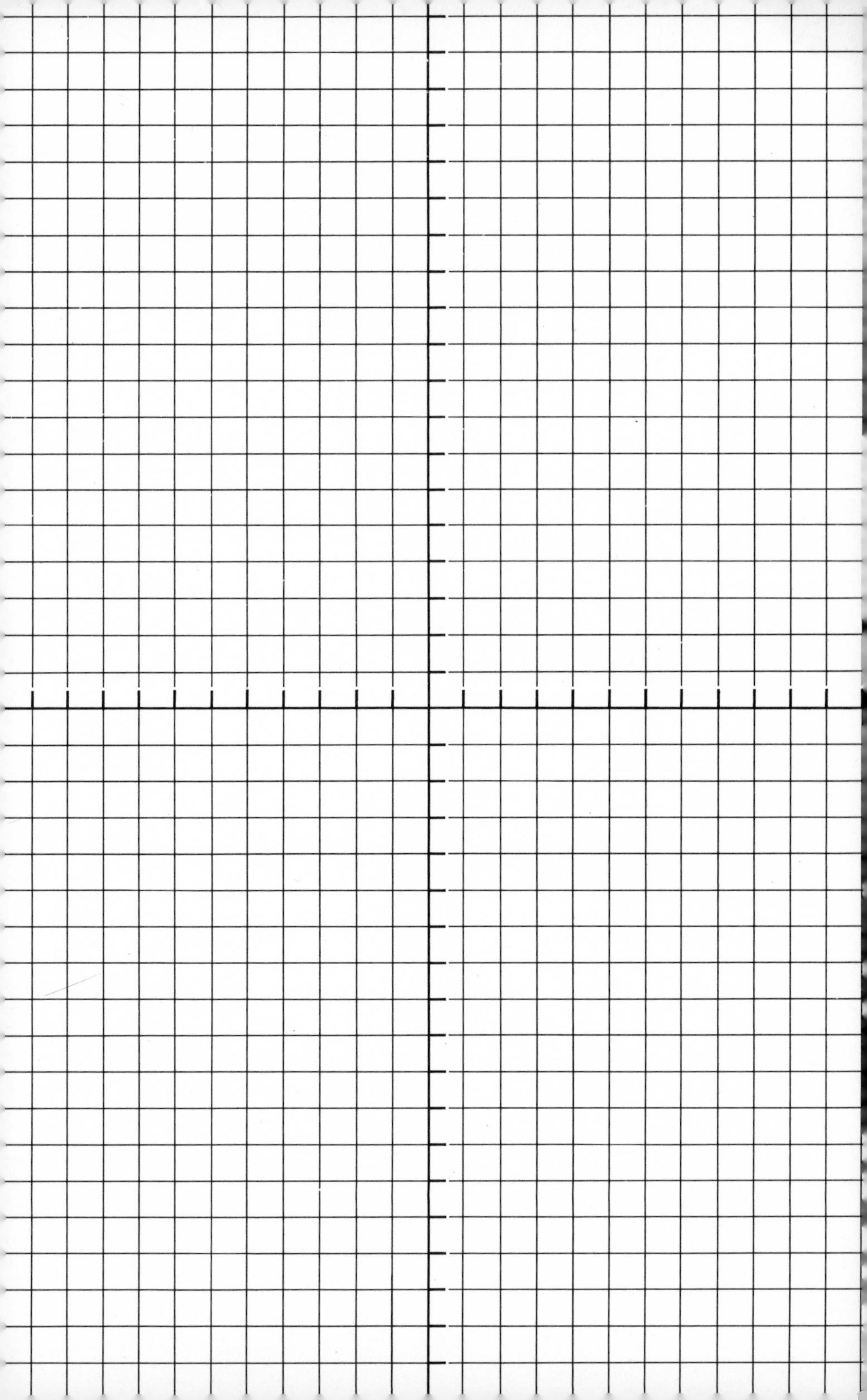

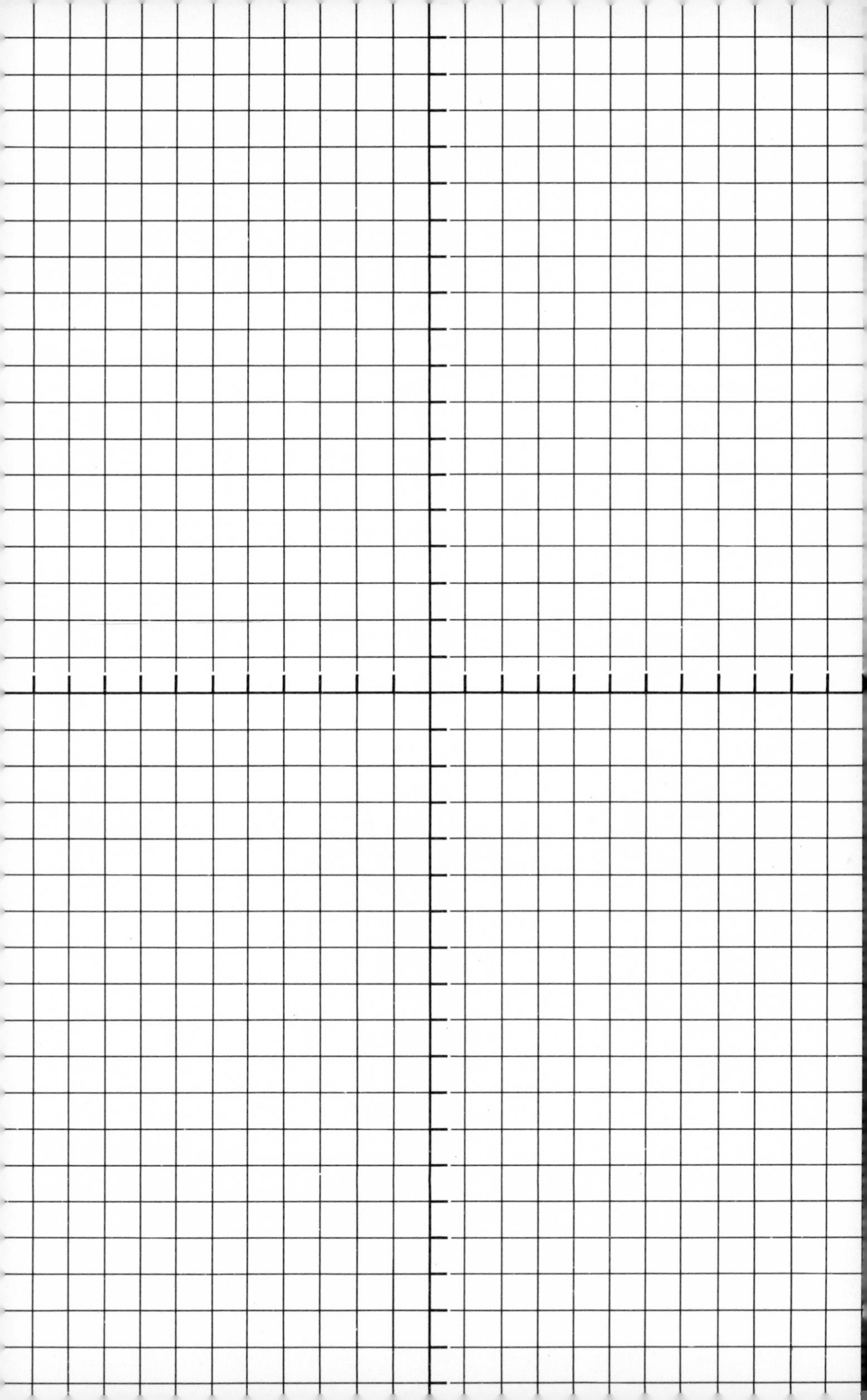